THE DIVINING EDGE

"LEARNER'S COMPREHENSIVE GUIDE TO DECODE GROUNDWATER DIVINING"

RUPESH DHANANJAI

ISBN
Paperback 979-8-89673-767-4
Hardcase 979-8-89906-298-8

"The Earth, the air, the land and the water are not an inheritance from our forefathers but on loan from our children. So we have to hand over to them at least as it was handed over to us."

—Mahatma Gandhi

This book is dedicated to my wife
Dr Anupama
and
my sons,
Anubhav and ***Abhinav***

Preface

Water is life—this is a truth as old as humanity itself. But as populations grow, climates shift, and resources dwindle, the challenge of finding and managing water has become one of the most pressing issues of our time. For engineers, farmers, diviners, water professionals, students, and NGOs working at the forefront of this challenge, the stakes couldn't be higher. *The Divining Edge* is a guide for those who seek to navigate these complexities, combining science, intuition, and practical tools to uncover Earth's most vital resource.

This book was born out of a deep fascination with the invisible forces that shape our landscapes—faults, fractures, and hydrogeomorphic patterns—and their critical role in determining where water flows, collects, and sustains life. Over the years, as I worked in fields both literal and figurative, I realised that bridging the gap between technical knowledge and on-the-ground action is essential for addressing water-related challenges in agriculture, urban planning, and community development.

The Divining Edge aims to empower readers and learners with tools like divining props, resistivity surveys, field-based lineament analysis, and hydrogeomorphic mapping—methods that combine cutting-edge science with hands-on divining practicality. But beyond techniques, this book is about perspective: seeing the land not just as soil and rock, but as a living, breathing system where water is both the thread and the lifeblood.

Whether you are an engineer designing sustainable infrastructure; a farmer seeking to optimise water use; diviners, learners, students, enthusiasts, water professionals, hydrologists working on resource management; or an NGO striving to bring clean water to underserved communities, this book is for you. Its goal is to provide not only technical knowledge but also a sense of wonder and respect for real diviners and the intricate systems that sustain life.

I invite you to explore these pages, learn from the insights shared, practice and apply them in your work. Together, we can move closer to a future where water is understood, respected, and available to all who need it. This is the journey of *The Divining Edge*—a journey of diviners' vision to see beneath the surface and uncover the vital connections that sustain our world.

Forward

In a world teetering on the brink of the unknown, the greatest discoveries often emerge from the sharpest edges—those moments where courage meets uncertainty, where vision clashes with reality, and where humanity dares to reach beyond its grasp.

The Divining Edge is not just a book; it is a journey into the profound. It calls upon the seeker, the dreamer, and the challenger within each of us to navigate the mysteries that shape our existence. It asks the questions we're too afraid to voice, explores the answers we're too hesitant to believe, and dares us to redefine what it means to thrive at the cusp of possibility.

As you turn these pages, let the edge guide you. Let it test your assumptions, ignite your curiosity, and unveil new perspectives. This book is an invitation—not to find comfort in the known, but to embrace the transformative power of the yet-to-be-discovered.

So, take a breath. Prepare yourself. And step to the edge.

The adventure awaits.

With best wishes,

Dt. 12.12.2024

Shri Rajesh Gupta,
**Chief Engineer, PHED
Raipur (Chhattisgarh)**

Acknowledgments

No journey, no matter how solitary it may feel at times, is ever truly walked alone. Writing *The Divining Edge* has been a profound experience, and I owe an immense debt of gratitude to those who have supported, inspired and encouraged me along the way.

To my family, who have been my rock and my sanctuary, thank you for your unwavering belief in me, even on the days when doubt loomed larger than hope. Your love and patience have been the foundation of my courage.

To my friends, mentors, and collaborators Shri P.S. Rana, hydrologist; Shri Manish Singh; Shri D. S. Rajput; Shri Yogesh Sinha; Shri Kishan Soni as the designer of sketches; and Shri Mithlesh Dahariya, thank you for your wisdom, your insights, and your willingness to challenge me to dig deeper. Your words of encouragement and constructive feedback have sharpened this book into what it is today.

To my readers, thank you for entrusting me with your time and imagination. Your curiosity and openness are the soul of this book, and I am honoured to share this journey with you.

To the unsung inspirations—the books, experiences, and moments that sparked the ideas within these pages—I am eternally grateful. You remind me that creativity is a constant dialogue with the world around us.

Finally, to the quiet force within me, thank you for whispering, "Keep going," when the path seemed unclear. Writing this book has taught me that The Divining Edge is not something to fear but something to embrace, to stand upon, and to dive from, into the infinite possibilities beyond.

This book is as much yours as it is mine. Thank you for being part of this journey.

Introduction

The global challenge of providing wholesome water to rural and underserved communities is one of the most pressing issues facing India and the world today. Access to safe and sustainable water is not just a matter of convenience—it's essential for health, economic development, education, and overall well-being. However, many rural and underserved communities face significant barriers in obtaining clean water. These challenges are multifaceted and often involve a combination of lack of infrastructure and economic, social, and environmental issues.

(A) Key Challenges

1. Geographical Barriers

- Many rural communities are located in remote, hard-to-reach areas with difficult terrain or are isolated by vast distances. This makes it more difficult and expensive to transport water or build infrastructure like wells, pipelines, and treatment facilities.

- In some areas, especially in arid or semi-arid regions, freshwater sources are scarce, and groundwater resources may be difficult to access or over-exploited.

2. Inadequate Infrastructure

- In many rural or underserved regions, existing water infrastructure is either outdated or nonexistent. Even where infrastructure is in place, it may be poorly maintained, leading to broken pipes, contamination, and unreliable water supplies.

- In some places, people rely on surface water sources (rivers, lakes, ponds), which are often contaminated with pollutants, pathogens, or heavy metals, making them unsafe for drinking without treatment.

3. Lack of Financial Resources

- Building and maintaining water infrastructure is expensive, and many rural communities lack the financial resources to invest in long-term solutions.

- Governmental and international aid may not be sufficient or consistently available, and private investments in water infrastructure can be limited, particularly when returns on investment in rural areas are perceived as low.

4. Water Pollution

- Pollution from agricultural runoff (pesticides, fertilisers), industrial waste, and untreated sewage can contaminate water sources, making it unsafe for consumption. In many rural areas, there is little to no wastewater treatment, which exacerbates contamination.

- Inadequate sanitation systems often result in faecal contamination of drinking water, leading to waterborne diseases like cholera, diarrhoea, and dysentery.

5. Climate Change

- Climate change is making the availability of water more unpredictable. Increased frequency of droughts, shifting rainfall patterns, and extreme weather events like floods and storms can disrupt local water supply systems.

- In some regions, glaciers that provide vital freshwater are melting, while in others, aquifers are being depleted faster than they can recharge due to overuse.

6. Cultural and Social Barriers

- In many rural communities, particularly in developing countries, women and children are often responsible for collecting water, which can be time-consuming and physically demanding. This can result in missed educational opportunities for children, particularly girls.

- Cultural and religious factors can also influence attitudes towards water management, sanitation, and hygiene practices, and may require sensitive, community-specific approaches to change behaviour and improve water access.

7. Lack of Education and Awareness

- A lack of knowledge about water conservation, sanitation, and hygiene can contribute to poor water management and the spread of waterborne diseases.

- Public education campaigns and community engagement are often needed to help people understand the importance of safe water, proper sanitation, and hygiene practices.

Despite the challenges, there are several innovative and practical solutions that can help address water access issues in rural and underserved communities. These solutions must be context-specific, sustainable, and inclusive.

(B) Solutions and Strategies

1. Improved Water Supply and Distribution Systems

- **Rainwater harvesting**: In areas with seasonal rainfall, rainwater harvesting systems can be an effective way to supplement water supply. These systems collect and store rainwater from rooftops or other surfaces, providing a clean and reliable source of water.

- **Solar-powered pumps**: Solar energy is being increasingly used to power water pumps and purification systems, particularly in remote regions where the electrical grid is absent or unreliable. Solar-powered pumps can lift groundwater or distribute water to remote villages.

- **Community-managed water systems**: Empowering local communities to manage and maintain their own water systems can improve sustainability. This might include local water committees or cooperatives that oversee the operation and upkeep of wells, pumps, or filtration systems.

2. Water Treatment and Purification

- **Point-of-use water treatment**: Simple, affordable technologies such as chlorine tablets, bio-sand filters, or solar disinfection (SODIS) can help purify contaminated water on a household level, improving water safety.

- **Filtration technologies**: In areas with heavy contamination, advanced filtration systems (e.g., reverse osmosis, activated carbon, UV sterilisation) can be used to remove pathogens, heavy metals, and other contaminants from water sources.

3. Sanitation and Hygiene Improvements

- Addressing sanitation alongside water access is essential. Simple, low-cost solutions like **pit latrines** or **eco-friendly toilets** can help reduce contamination of water sources.

- **Improved hygiene education** can teach communities about handwashing, safe water storage, and proper waste disposal, reducing the incidence of waterborne diseases.

4. Financial Innovations and Public-Private Partnerships

- **Microfinance** and **community-based financing** models can provide rural communities with the resources needed to build and maintain their water infrastructure. These models enable local ownership and long-term sustainability.

- **Public-private partnerships**: Governments, NGOs, and private companies can work together to develop and fund water projects. The private sector can bring technical expertise and financing, while governments and NGOs can ensure that the solutions are equitable and inclusive.

5. Advancing Data and Monitoring

- **Geospatial technologies** like satellite imagery and geographic information systems (GIS) are being used to map water resources, predict water availability, and identify areas at risk of water scarcity or contamination.

- **Water quality monitoring**: Regular testing of water sources for contaminants can help ensure that communities have access to safe drinking water.

6. Climate Resilience and Adaptation

- **Drought-resistant crops** and **water-efficient irrigation systems** can help rural communities adapt to changes in rainfall patterns caused by climate change.

- Building resilience through **climate-smart water management** can ensure that communities are better prepared for changing water availability.

◆ ◆ ◆

(C) Global Efforts and Initiatives

In this regard, global efforts and initiatives are ongoing as follows:

1. **United Nations Sustainable Development Goal (SDG) 6:** The UN has set a global target to "ensure availability and sustainable management of water and sanitation for all" by 2030. This includes improving water quality, increasing water-use efficiency, and expanding water and sanitation access to rural and underserved areas.

2. **World Bank and Global Water Partnerships:** The World Bank and other development organisations, like the Global Water Partnership (GWP), are working to provide financing and technical assistance for water infrastructure projects in rural and underserved areas.

3. **Community-Led Total Sanitation (CLTS):** This is an approach that empowers communities to take responsibility for improving sanitation and water access. By focusing on community participation and local leadership, CLTS has been successful in many areas, particularly in Africa and South Asia.

4. **NGOs and grassroots organisations:** Many NGOs, such as **WaterAid, The Water Project,** and **charity: water,** focus on

providing sustainable water solutions to rural and underserved populations. They often work with local communities to ensure that solutions are culturally appropriate and sustainable.

Engineering plays a critical role in solving the complex water challenges faced by rural and underserved communities. Traditional engineering solutions, particularly in the areas of **borehole drilling**, **hydrology**, and **water management**, have been central to accessing, managing, and distributing water in these regions. While engineering solutions are often combined with modern technologies, many traditional methods remain effective, practical, and adaptable to local conditions.

1. **Borehole drillings** are one of the most common methods used to access groundwater, especially in rural and arid areas where surface water is limited. The role of engineering in borehole drilling is pivotal in ensuring safe, sustainable, and effective access to water.

2. **Hydrology**To understand water resources, **hydrology** is the branch of science and engineering that deals with the distribution, movement, and quality of water resources in the environment. In rural and underserved communities, understanding local hydrological conditions is essential for ensuring that water resources are used sustainably and efficiently.

3. **Water management** engineering involves organising the use, distribution, and conservation of water resources to meet the needs of communities while ensuring sustainability. It encompasses the planning, design, operation, and maintenance of water systems for domestic, agricultural, and industrial purposes.

   Traditional engineering solutions in borehole drilling, hydrology, and water management have been essential in helping rural and underserved communities access and manage water. While modern technologies like solar pumps, advanced filtration, and automated monitoring systems are increasingly being adopted, traditional methods remain relevant and effective in many regions.

In the future, successful water management will likely involve a combination of **traditional engineering** and **modern technologies**—especially in developing countries or remote regions where infrastructure and financial resources may be limited. Collaboration between engineers, local communities, and policymakers is essential to develop sustainable, context-specific solutions that ensure safe and reliable water access for all.

Groundwater divining is a fascinating blend of history, folklore, and practical application. While its roots are deeply embedded in mystical traditions, it remains a testament to humanity's ongoing quest to connect with the unseen forces of the natural world. Whether viewed as a superstition or a legitimate practice, divining continues to intrigue and inspire people around the globe.

Different cultures have developed their own methods and beliefs surrounding dowsing:

- **Europe**: In Germany and parts of France, dowsing with a forked hazel stick was widely practised. The dowser would hold the stick at each end, and the fork would bend or twitch when it passed over a water source.

- **China**: Early Chinese records mention the use of dowsing rods for locating minerals and water. Chinese diviners often used a single, Y-shaped branch.

- **The Americas**: In some Native American cultures, dowsing was linked to ritualistic practices, connecting it to the spiritual world.

- **India:** In many parts of the nation, divining practices combine with human intuitions, like baiga's mantras, coconut witching, rituals, etc.

The objective of this book is to explore how **traditional divining techniques** can complement **modern engineering practices** to create innovative, sustainable and cost-effective solutions for water resource management. By examining real-life case studies, historical practices and modern adaptations, this book seeks to bridge the gap between time-tested indigenous knowledge and advanced scientific methods.

This book aspires to inspire professionals, researchers, and policymakers to explore innovative, inclusive methods for managing the world's most precious resource—water.

◆ ◆ ◆

Contents

PART-II
(LEGIT GOALS)

- **Mind Maps 1 & 2**
- **Glossary of Terms**
  - Key terminology and definitions.
- **Resources and References**
  - Recommended reading.
  - Online resources and tools.

PART-I

OPTIMIZED METHODOLOGY

Chapter 1

Understanding Groundwater Divining: A Historical and Cultural Perspective

1.1. What is Groundwater Divining?

- Historical origins and cultural significance of divining for water in various regions (e.g., Europe, Africa, Asia).

- Different tools and techniques used (Y-rod, L-Rod, pendulum, etc.).

1.2. Divining and Indigenous Knowledge: The role of local communities in preserving and transmitting traditional knowledge.

1.3. Case Studies of Successful Divining: Highlight notable examples where divining has been used effectively in water sourcing, particularly in rural or resource-poor regions.

◆ ◆ ◆

1.1. What is Groundwater Divining?

Groundwater divining, also known as **dowsing** or **water witching**, is the practice of using a tool, typically a forked stick or rods, to locate underground water sources. Dowsers (or diviners) claim to be able to detect the presence of water, minerals, or other hidden substances beneath the Earth's surface by observing the subtle movements or vibrations of their tools in response to unseen forces.

This practice is often considered a blend of art, intuition, and tradition, with practitioners either using simple tools like a Y-shaped branch, pendulum, or dowsing rods, or relying on their "sensitive" abilities to detect the flow of groundwater.

1.1.1. Historical Origins of Groundwater Divining

The origins of divining for water date back thousands of years. While the exact origin of dowsing is difficult to trace, evidence of its practice appears in ancient cultures across Europe, Asia, and Africa. It has been used in various forms throughout history to locate water, minerals, and sometimes even treasure. The tools and methods may vary by region, but the fundamental principle—divining hidden forces or elements—remains the same.

Europe

- **Ancient Europe:** The practice of dowsing for water can be traced back to ancient civilisations, including the Egyptians, Greeks, and Romans. In these societies, divining for water was often linked with mysticism and spiritual practices. The use of a forked stick, or «Y-rod», is often cited as one of the earliest known tools used by European dowsers.

- **Medieval and Renaissance Europe:** In medieval Europe, dowsing became a widely accepted practice, particularly for locating underground water sources. People would use divining rods, typically made of hazel wood or willow, to find wells and springs. By the 16th century, the practice had become so common that even physicians and scholars like Paracelsus (a Swiss physician) acknowledged it, though not without controversy. The practice was sometimes associated with witchcraft, and practitioners were known as «water witches» due to the mystical associations of the craft.

- **Modern Europe**: In more recent times, dowsing has persisted, especially in rural communities where access to advanced technology is limited. It has also been a part of the folklore and traditions in countries like Germany, France, and England. The use of rods for water locating remains a popular and traditional practice in some parts of Europe, even today.

Africa

- **West Africa**: In many West African cultures, dowsing is closely tied to spiritual practices. The belief in mystical forces guiding the dowser is widespread, and it is common to find individuals who are regarded as possessing special gifts or spiritual insight to locate water and resources. In rural African communities, where access to modern water surveying technologies can be limited, dowsing is still used as a tool to find reliable water sources for wells.

- **Africa**: The practice of «water witching» or dowsing is also found in Southern Africa, where it is often seen as a way to communicate with the spiritual world. In countries like Zimbabwe and South Africa, dowsers use a variety of methods, such as divining rods, to find underground water, and it is often connected to traditional spiritual beliefs and practices.

Asia

- **China**: In ancient China, divining for water, minerals, and other resources was practised with great reverence. The earliest known references to dowsing in China date back to the 3rd century BCE when it was associated with Taoist practices of divination. Dowsers in China typically used a Y-shaped branch or a rod made from a single piece of wood, similar to practices in other cultures, and the technique was closely linked to the concept of Qi, or vital life force. Chinese diviners believed that the flow of Qi could be influenced by the presence of water or minerals underground, and dowsing was seen as a way to tap into this invisible force.

- **India**: In India, traditional water divining practices have also existed for centuries. Hindu scriptures mention the use of divining rods and other tools to locate water, and the practice was often associated with the sacred and spiritual realm. In some rural parts of India, dowsing continues to be used in combination with more modern methods to find underground water, especially in arid regions.

The Americas

- **North America**: Indigenous peoples in North America did not widely practice dowsing in the same way as Europeans, but there are records of certain Native American tribes using divining methods to locate water sources. The practice of water divining was later adopted by settlers and pioneers in the 17th and 18th centuries, particularly for finding wells in the American frontier.

- **South America**: In parts of South America, particularly in rural areas, dowsing was practised by early settlers and indigenous people alike. It was often combined with shamanistic beliefs and rituals that involved communication with the natural world to locate water or other valuable resources.

1.1.1.a. Cultural Significance of Divining for Water

The act of divining for water carries deep cultural significance in many regions. In most cases, it is not merely a practical tool but a spiritual practice that connects people to the natural world in profound ways.

1. **Mysticism and Spirituality**: Dowsing is often associated with mystical beliefs about the hidden forces of nature. In many cultures, the dowser is seen as someone with special sensitivity or spiritual gifts, who is able to connect with the Earth or spirit world. For example, in parts of Africa and Asia, dowsing may be tied to ancestral spirits or the belief in a spiritual connection to the land.

2. **Tradition and Folklore**: In many regions, especially in rural areas, dowsing is a cultural tradition passed down through generations. The tools used for dowsing, such as branches, rods, or pendulums, often have symbolic importance. In some European cultures, for example, the use of hazel or willow wood for divining rods is rooted in ancient symbolic associations with fertility and life-giving forces, as these trees were believed to be sacred.

3. **Practical Use**: Despite scepticism and the rise of modern scientific methods, groundwater divining remains a practical tool in many parts of the world. In areas where water scarcity is a problem, such as arid regions in Africa, or where modern survey equipment is too expensive or unavailable, dowsing can provide a simple way to locate underground water sources. Though its scientific validity is debated, many still

rely on dowsing as a supplementary tool in water resource management.

1.1.2. Different Tools and Techniques Used in Groundwater Dowsing

Groundwater divining (or dowsing) involves the use of various tools that practitioners believe can help detect the presence of water, minerals, or other hidden substances beneath the ground. While different cultures and traditions have developed unique methods over the centuries, the core idea behind all of them is that these tools react to unseen forces—such as the flow of water or energy fields in the Earth—when passed over a source.

Chapter six details **the common tools and techniques used in divining: The tools and techniques used in groundwater divining vary widely depending on the culture, tradition, and personal preference of the dowser. Whether using the iconic Y-rod, the more precise L-Rod, the intuitive pendulum, coconut, straight zinc or copper rods**, each method in detail is given in chapter 6, all rely on the belief that the practitioner can tap into unseen natural forces. While the scientific community remains sceptical about the efficacy of these methods, many still swear by them for locating groundwater, minerals, and other hidden resources. For those who practice divining, these tools provide a deeply personal and often spiritual connection to the natural world.

1.2. Divining and Indigenous Knowledge: The Role of Local Communities in Preserving and Transmitting Traditional Knowledge

Groundwater divining (or dowsing) is not just a practice rooted in mystical or scientific tradition; for many Indigenous communities around the world, it is part of a broader system of knowledge that connects people to the land, water, and spiritual forces. The role of divining in these communities goes beyond merely locating underground water—it serves as a way of understanding and interacting with the natural environment. In this context, divining is not only about practical outcomes but also about preserving cultural identity, maintaining ecological balance, and passing down important knowledge through generations.

Indigenous Knowledge Systems: A Holistic Approach

Indigenous knowledge systems are characterised by their **holistic** view of the world, in which human beings, nature, and spiritual forces are deeply interconnected. For many Indigenous cultures, traditional knowledge is transmitted orally and is deeply tied to a place—its history, stories, and natural features. Water, in particular, holds significant symbolic and practical importance in these knowledge systems.

In the case of divining, the practice can be understood not just as a method for finding water, but as part of a broader relationship with the environment. The practice of water divining is often viewed as a sacred act, one that involves more than just physical tools or movements. It requires **spiritual sensitivity**, intuition, and the ability to read the land and its energies. In many Indigenous cultures, water is seen as a living force—often personified—and those who can locate it are seen as having a special connection with this life-giving element.

1.2.1. Role of Divining in Indigenous Communities

1. Preserving Ecological Knowledge

For many Indigenous communities, especially in rural or remote areas, **water** is a scarce and vital resource. Dowsing has historically played an important role in helping these communities locate underground water sources, especially in regions where modern technology or infrastructure might be limited. **Traditional dowsers** use their knowledge of local geography, plant life, and the environment, often guided by spiritual beliefs or ancestral wisdom.

In this context, divining is not just a mystical or esoteric practice but an essential tool for **sustainable living**. By identifying water sources, local communities ensure access to life-sustaining resources. This knowledge of where water is found, how to access it, and how to maintain its purity is passed down through generations, contributing to the survival and resilience of communities in challenging environments.

2. Connection to Spiritual and Cultural Identity

Divining is often tied to **spiritual practices** in many Indigenous cultures, with diviners viewed as intermediaries between the physical world and the spiritual realm. Water is frequently seen as sacred, and the ability to locate it can carry deep symbolic meaning, connected to concepts of fertility, life, and ancestral continuity.

For example, among certain Indigenous groups, the practice of divining is linked to the reverence of **water spirits** or **ancestral entities** who are believed to govern or protect water sources. The act of locating water through dowsing is therefore not just about practical outcomes, but also about maintaining **harmony** with the spiritual forces that govern the land.

In many Indigenous cultures, **divining rituals** are often accompanied by offerings, prayers, or ceremonies, reinforcing the idea that the search for water is a sacred task. For these communities, passing down this knowledge is essential for maintaining their connection to the land and the spiritual dimensions of their world.

3. Intergenerational Transmission of Knowledge

The practice of divining is often passed down orally, from **elders to younger generations**, as part of a larger system of cultural transmission. Elders serve as both **knowledge keepers** and **spiritual guides**, teaching younger community members not only how to use divining tools but also how to interpret subtle environmental signs, such as animal behaviour, plant growth, and geological features, that indicate the presence of water.

In this way, the practice of divining is deeply tied to the wider **cultural education** of the community. Diviners do not simply teach the technical aspects of the practice (e.g., how to use rods or sticks), but also the **ethical and spiritual principles** behind it. This includes an understanding of why water is sacred, how to protect it, and how to approach the land with respect.

For example:

- In **West Africa**, some communities use divining to find water and ensure the health of their wells, often accompanied by community rituals.

- In parts of **South America**, Indigenous groups use divining as part of their **cosmological understanding** of the Earth and water as living beings that must be respected and nurtured.

The knowledge of **when** and **where** to look for water is carefully guarded and passed down, often through specific family lines or clans, ensuring that only those with proper training and respect for the practice are entrusted with the responsibility.

4. Maintaining Cultural Resilience and Autonomy

For many Indigenous peoples, **cultural resilience** is intertwined with the ability to maintain traditional practices and knowledge, even in the face of modernising forces and outside pressures. As communities confront challenges such as climate change, deforestation, or land displacement, traditional knowledge systems like divining can offer ways to adapt and survive.

In this context, divining is not just about the technical ability to find water; it is also about maintaining **cultural autonomy**. In regions where Western technological solutions may not be available, or where traditional land rights and resources are threatened, the ability to **locate water** through divining can help communities **assert their sovereignty** over the land and its resources.

In some cases, the use of divining may even be a form of **resistance** to modern agricultural or industrial practices that threaten traditional ways of life. By continuing to practice divining and other traditional ecological knowledge, Indigenous communities are preserving alternative ways of engaging with the environment, which can offer sustainable solutions to modern crises.

◆ ◆ ◆

1.2.2. The Role of Local Communities in Preserving Divining Knowledge

Indigenous communities are at the forefront of preserving and transmitting the knowledge of divining, often despite external challenges. Several factors contribute to the effective preservation of this knowledge:

1. Oral Tradition and Storytelling

Oral traditions are vital to the transmission of knowledge, particularly in Indigenous cultures where writing systems may not be the primary means of recording history. Elders pass down the practice of divining through stories, teachings, and hands-on experiences, ensuring that knowledge is rooted in lived experience and spiritual understanding.

2. Cultural Revitalisation Movements

In some regions, there are active efforts to **revitalise traditional practices** like divining. These efforts are part of broader movements to reclaim and celebrate Indigenous knowledge systems, particularly in the face of globalisation, colonialism, and the erosion of traditional ways of life. Divining, as part of this effort, becomes both a **symbol of resistance** and a practical means of maintaining cultural continuity.

3. Intergenerational Mentorship

As in many traditional knowledge systems, **mentorship** plays a critical role in the transmission of divining practices. Younger generations learn from experienced practitioners, often over many years, developing not only the technical skills to use the tools but also the wisdom to interpret the natural world with respect and understanding. This mentoring relationship ensures the continued relevance and integrity of the practice.

1.3. Case Studies-1 for Successful Divining: Effective Use in Water Sourcing

Here are some notable **case studies** where dowsing has been successfully used in water sourcing:

1. West Africa: The Use of Divination in Burkina Faso

Context

In parts of **Burkina Faso**, particularly in rural areas where the climate is dry and water scarcity is a constant challenge, traditional methods of locating groundwater are still widely used. The region faces recurring droughts, and many communities rely on **deep wells** for water. The country has a significant **rural population**, with limited access to modern water-sourcing technologies, particularly in remote areas.

Use of Divining

In Burkina Faso, **local dowsers** (known as **"water witches"**) use traditional divining techniques to locate underground water sources. The tools used include **Y-shaped sticks** (typically made from **hazel wood**) or **metal rods**, similar to the methods used in Europe. These diviners often work alongside **hydrogeologists** and **community members** to pinpoint potential water-bearing areas for wells, particularly in regions that are difficult to survey with modern equipment.

Success

- **Community reliance** on diviners has proven effective in areas where no modern geological surveys or drilling equipment are available.

- In one rural community, a diviner was able to identify a **reliable water source** under a plot of land that had previously been surveyed by technicians without success. Upon drilling, the borehole yielded a sufficient and consistent supply of groundwater.

- The combination of traditional divining with **modern drilling techniques** has increased the success rate of water projects in areas where resources are scarce. **Divining** helps reduce the guesswork involved in locating viable drilling sites, thus cutting down on wasted time and resources.

Cultural Significance

- The practice is deeply embedded in the community's **spiritual and cultural practices**, as the role of the diviner often carries with it a sense of **ancestral authority**. The belief that divining is not only a practical skill but also a spiritual calling reinforces its importance in the region.

- Diviners are seen as having an intimate connection with the land, and their guidance in finding water is often sought for other resources, such as mineral deposits.

◆ ◆ ◆

2. South Africa: Water Witching in the Karoo

Context

The **Karoo** region in **South Africa** is an arid and semi-arid area characterised by low rainfall and limited access to groundwater. This vast area is home to a number of rural communities that face significant water challenges. Many of these areas are **resource-poor** and are difficult to access with modern water detection technologies due to the **hard rock formations** that complicate traditional geological surveys.

Use of Divining

In South Africa, particularly in the **Karoo**, water witching using **divining rods** is a common practice. Local dowsers, who are often experienced in reading the land and its natural signs, use **copper or zinc rods** to locate underground water. These rods are believed to react to the presence of water when held by the practitioner and moved across the land.

Success

- In one particular case in the Karoo, a family that had been struggling to find water for their farm turned to a **local dowser**. The diviner used the rods to locate water beneath a rocky area where earlier geophysical surveys had failed. The drilling at the site yielded a **strong and reliable water source** that provided water for both domestic use and irrigation.

- This success led the community to seek out other dowsers, and water witching became a more trusted method in the region, particularly in the **absence of modern hydrological expertise**.

- Water witching has been used successfully in rural areas for locating water for **small-scale farming**, livestock, and even domestic needs. In areas where traditional geological mapping has been costly or ineffective, divining provides a low-cost alternative for sourcing water.

Cultural Significance

- Divining in the Karoo is culturally significant, with many local practitioners believing they are following a **spiritual tradition** passed down through generations. The ability to find water is seen as a **gift**, and successful diviners are respected members of their communities.

- In the context of **water scarcity**, the practice has helped maintain **local autonomy** over resources, allowing rural communities to manage their water needs independently.

◆ ◆ ◆

3. India: Traditional Water Sourcing in Rajasthan

In Rajasthan, wells and rainwater harvesting are essential to meet the water needs of the people, as the region has a significant reliance on traditional knowledge and indigenous water management practices due to the lack of modern infrastructure and the arid climate.

Use of Divining

In Rajasthan, water diviners use a combination of **rods** and **sticks** to locate underground water sources. The practice is often referred to as **"jaljivan"** (meaning "water life" in Hindi), and it has been passed down for generations. Diviners (sometimes called **"water prophets"**) walk across the land, using either **metal rods** or **wooden sticks**, to sense the presence of groundwater.

Success

- In several parts of Rajasthan, local diviners have been instrumental in identifying **water locations** for the construction of wells in rural villages. One prominent case involved a village in the **desert region** that had been struggling with water shortages. A local dowser successfully located water under a site that had previously been dismissed by engineers, and drilling revealed a **sizeable aquifer** that provided the community with enough water for drinking and agriculture.

- In another instance, a water diviner helped locate a **water source for an irrigation well**, enabling a community to continue growing crops during a particularly severe drought. The successful identification of underground water allowed for **sustainable farming practices** in a region that often faces the threat of food insecurity due to water scarcity.

Cultural Significance

- The practice of divining for water in Rajasthan is tied to **spiritual beliefs**, where diviners are sometimes viewed as **spiritually gifted** individuals with the ability to commune with the Earth. The act of divining is often accompanied by **rituals** or prayers, and diviners are highly respected members of the community.

- As **climate change** and water scarcity threaten traditional ways of life, the continued use of water divining helps preserve a vital cultural heritage while contributing to the **sustainability of local communities.**

◆ ◆ ◆

4. Latin America: Water Divining in Mexico and Peru

Context

In several parts of **Latin America**, including **Mexico** and **Peru**, access to clean water is a critical issue, particularly in **rural communities** and **mountainous regions**. Many of these areas are difficult to reach with modern geological surveying equipment, and rural residents often rely on **hand-dug wells** to access groundwater.

Use of Divining

In these regions, dowsing has been used for centuries as a way to locate underground water in areas where traditional surveying tools may be ineffective or unavailable. **Local dowsers** often use **metal rods** or **Y-shaped sticks** to detect the presence of water beneath the surface.

Success

- In **Mexico**, for example, a community in the **mountainous regions** of the state of **Chihuahua** used divining to locate water during a severe drought. The local diviner was able to find a **water-bearing vein** beneath a large rock formation, which was then successfully drilled into, providing the community with a much-needed water source.

- In **Peru**, particularly in the **Andean highlands**, water divining has been used to find water sources for both **drinking** and **irrigation**. In these remote regions, where modern water technologies are less accessible, divining is seen as a cost-effective and culturally relevant method to locate water, and it has supported sustainable farming practices in communities that face extreme climate conditions.

Cultural Significance

- In both Mexico and Peru, the practice of divining is embedded in **traditional knowledge** that combines **spiritual** and

ecological wisdom. Water diviners are often regarded as people holding special **wisdom** about the land, and their role extends beyond finding water to maintaining **balance** with the Earth and respecting sacred **water spirits**.

- These practices reinforce the community's **relationship with the land**, which is viewed as a living entity deserving of respect.

1.4. Conclusion

Groundwater divining, with its deep historical and cultural roots, remains a fascinating blend of practicality and mysticism. Across various cultures—from ancient civilisations in Europe and Asia to African traditions and rural communities in the Americas—divining for water has been a vital and spiritually significant practice. Though the practice has faced scepticism, it continues to be valued in many parts of the world for its perceived ability to tap into the invisible forces of nature, connecting humans with the Earth in unique and powerful ways.

Divining for water is a practice that, for many Indigenous communities, is not merely a technique for finding underground resources, but a profound part of a larger system of ecological, spiritual, and cultural knowledge. This knowledge passed down through generations, helps maintain sustainable relationships with the environment, fosters spiritual connections to water and land, and strengthens community resilience.

For local communities, preserving and transmitting traditional knowledge like divining is an act of cultural survival and autonomy. As the world faces increasing challenges related to climate change, water scarcity, and environmental degradation, Indigenous practices like divining offer valuable insights into sustainable living and the deep wisdom embedded in centuries of interaction with the Earth.

By maintaining these traditions, Indigenous communities not only honour their ancestors but also ensure that future generations can continue to live in harmony with the natural world.

These case studies illustrate the continued relevance and effectiveness of divining for water sourcing in rural and resource-poor regions around the world. In communities facing water scarcity, especially where modern technologies are unavailable or impractical, **traditional divining practices** offer a reliable and cost-effective alternative. While the scientific community may question the mechanisms behind these practices, the success stories in places like **West Africa, South Africa, India**, and **Latin America** underscore the **practical value** of traditional knowledge systems in addressing contemporary water challenges.

By integrating traditional practices like divining with modern technologies and knowledge, these communities are able to maintain **self-sufficiency, sustainability,** and **cultural continuity,** ensuring access to groundwater.

Chapter 2

The Science of Groundwater and Water Resource Planning

2.1. Hydrology and Groundwater Systems: A brief overview of groundwater science and hydrological cycles.

- Aquifers, borewells, groundwater flow.

- Limitations of traditional methods of locating water sources.

2.2. Engineering Practices in Borewell Drilling:

- Traditional vs. modern drilling methods.

- The role of geophysical surveys, hydrogeological mapping, and other scientific tools in determining water locations.

2.3. The Gap in Traditional Engineering:

- Why traditional engineering methods might not always be successful in rural and developing areas (lack of access to advanced technology, terrain challenges, cost).

◆ ◆ ◆

2.1. Hydrology and Groundwater Systems: An Overview

Hydrology is the study of water in the environment, particularly its distribution, movement, and properties within the Earth's systems. Groundwater, one of the most critical sources of fresh water worldwide, is an essential component of hydrological systems. Understanding how groundwater behaves and flows is crucial for the management of water resources, especially in regions where surface water is scarce or unreliable.

This section will provide a brief overview of the **hydrological cycle**, **aquifers**, **boreholes**, and **groundwater flow**, along with the **limitations of traditional methods** for locating groundwater sources.

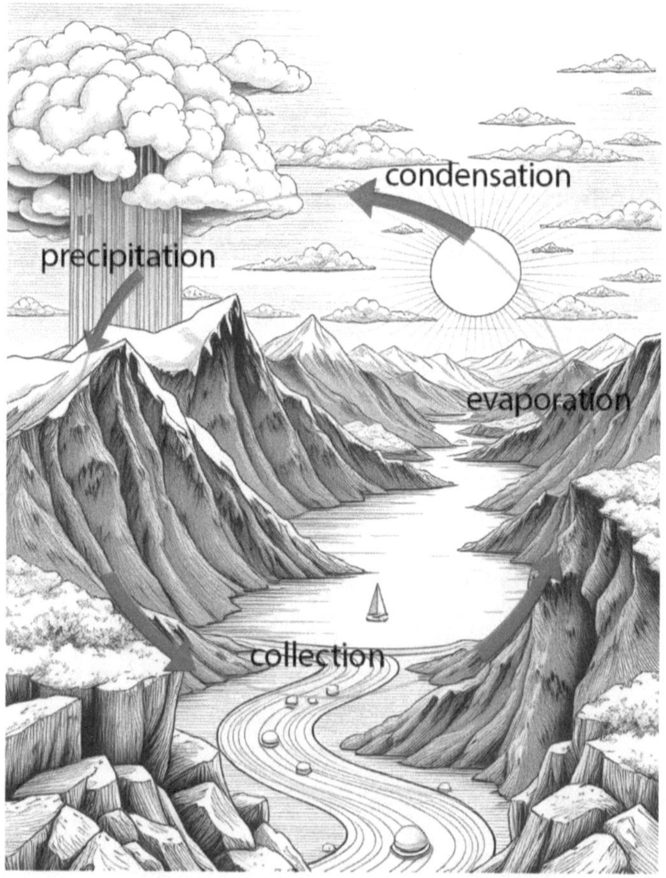

Fig-1 The Hydrological Cycle(Illustration map)

2.1.1. The Hydrological Cycle

The **hydrological cycle**, or **water cycle**, is the continuous movement of water within the Earth's atmosphere and surface. The cycle describes how water moves between the **atmosphere, land**, and **ocean** through various processes such as **evaporation, condensation, precipitation,** and **infiltration**. Groundwater is a critical component of this cycle, as it plays a role in storing water that infiltrates into the ground.

Key Phases of the Hydrological Cycle

- **Evaporation**: Water from oceans, lakes, and rivers evaporates into the atmosphere.

- **Condensation**: As the water vapour rises, it cools and condenses into clouds.

- **Precipitation**: Water falls from clouds as rain, snow, or other forms of precipitation.

- **Infiltration**: Some of this water seeps into the ground, replenishing groundwater reserves.

- **Percolation**: Water that infiltrates into the ground moves downward through soil and rock layers, eventually reaching the water table, where it accumulates as groundwater.

- **Runoff**: The rest of the water moves across the land surface and returns to bodies of water like rivers, lakes, and oceans.

2.1.2. Aquifers and Groundwater Storage

An **aquifer** is a body of rock or sediment that can store and transmit groundwater. Aquifers are critical water sources for many regions, as they can provide a reliable supply of fresh water for drinking, irrigation, and industrial use.

a. Types of Aquifers

- **Unconfined Aquifer**: This type of aquifer is directly connected to the surface with the water table serving as the upper boundary. Water in unconfined aquifers can easily infiltrate from the surface through **precipitation** or **runoff**.

- **Confined Aquifer**: A confined aquifer is bounded above and below by impermeable layers, such as clay or rock, which restrict water movement. The water in confined aquifers is under pressure, and tapping into them often results in **artesian wells**, where water can flow to the surface without pumping.

- **Semi-confined Aquifer**: These aquifers are bounded by layers that are somewhat permeable, allowing some water movement but not as freely as unconfined aquifers.

Aquifers are typically replenished by **recharge zones**, where water from precipitation or surface water infiltrates into the ground, moving down to the water table. Groundwater flow within aquifers is influenced by the permeability and porosity of the rocks and sediments involved. **Permeability** refers to how easily water can flow through a material, while **porosity** refers to the amount of space within the material to store water.

b. Groundwater Flow

- **Flow Direction**: Groundwater typically flows from areas of **higher pressure** or **higher elevation** to areas of **lower pressure** or **lower elevation**. The flow is generally slow and can be influenced by geological formations, natural barriers, and human interventions.

- **Discharge**: Groundwater discharges into rivers, lakes, or wetlands, and sometimes flows to the surface through natural springs.

- **Recharge**: Recharge occurs when water from precipitation or surface water infiltrates into the ground and replenishes the aquifer. In some cases, recharge can be affected by human activities, such as over-extraction or land use changes.

2.1.3. Boreholes and Groundwater Extraction

A **borehole** is a deep, vertical hole drilled into the ground to access groundwater from aquifers. Boreholes are typically used for extracting water in **rural** and **remote areas**, where surface water resources (like rivers and lakes) are scarce or unreliable. The extraction can occur in both **unconfined** and **confined aquifers**, depending on the depth and geological formation.

a. Borehole Construction

- **Drilling**: A borewell is drilled using specialised equipment, often down to a depth where groundwater is found. The depth can range from a few metres to hundreds of metres, depending on the location and the depth of the water table.

- **Casing**: A casing (usually made of steel or PVC) is inserted into the borewell to prevent collapse and to protect the water from contamination.

- **Pump Installation**: Once the borewell reaches an aquifer, a pump is installed to extract the water.

b. Challenges with Boreholes

- **Depletion**: Over-extraction of groundwater can lead to the **depletion** of aquifers, causing a **lowering of the water table**. In some areas, this has led to the **drying up** of borewells, especially during periods of prolonged drought.

- **Contamination**: Borewells can also be susceptible to contamination from pollutants on the surface, especially if proper casing and sealing are not maintained.

- **Sustainability**: In many areas, the rate of groundwater extraction can outpace the natural recharge rate of aquifers, leading to unsustainable water use.

2.1.4. Limitations of Traditional Methods for Locating Water Sources

While traditional methods of locating water sources, such as **dowsing** or **divining**, have been used for centuries and are deeply embedded in various cultures, they have limitations when compared to modern hydrological techniques.

a. Lack of Scientific Basis

Traditional methods like dowsing rely on the idea that divining rods or other tools can "sense" water beneath the Earth. However, these practices are generally not grounded in modern **hydrology** or **geology**, and their effectiveness is often anecdotal or based on **cultural belief systems** rather than empirical evidence. The lack of a scientific explanation makes it difficult to assess the reliability of these methods in more technical or large-scale applications.

b. Inability to Assess Water Quality

Traditional divining methods focus on identifying the **presence** of groundwater, but they typically cannot provide detailed information about **water quality**. For instance, modern techniques like **hydrogeological surveys** or **borehole drilling** can assess factors like **salinity**, **contaminants**, and **the sustainability of the water source**, which are crucial for safe drinking water and long-term use.

c. Limited Depth Precision

Divining tools may help locate **general areas** of groundwater, but they often cannot pinpoint specific **depths** or **flow rates** of water accurately. **Hydrological surveys** and **geophysical methods** (such as

seismic surveys or **electrical resistivity** testing) can provide detailed, **quantitative information** about the depth, extent, and flow of aquifers, which is crucial for successful borehole drilling.

d. Geological Complexity

Groundwater systems are highly influenced by the **geological characteristics** of the region, including the type of rock, soil permeability, and the structure of the aquifer. Traditional methods may not account for **geological complexity**, such as the presence of **faults**, **layers of impermeable rock**, or **confining layers** in aquifers. Modern techniques like **remote sensing** or **ground-penetrating radar (GPR)** can provide valuable insights into these geological factors and guide more accurate drilling efforts.

e. Risk of Overuse

Traditional methods might lead to the discovery of water sources, but without an understanding of the **aquifer's capacity** or recharge rate, there is a risk of **over-extraction**, which can lead to **aquifer depletion** or **land subsidence**. Modern hydrological studies help assess the sustainability of water extraction and ensure that water sources are used in a way that does not degrade the environment.

2.2. Engineering Practices in Borehole Drilling: Traditional vs Modern Methods

Borehole drilling is a critical process for accessing groundwater, particularly in areas where surface water is scarce. Over time, the techniques used for borehole drilling have evolved from **traditional methods** to more advanced and scientifically guided **modern practices**. While traditional methods still hold cultural and practical significance in many parts of the world, modern engineering and scientific advancements have led to more precise, efficient, and sustainable borehole drilling practices. This section will compare

traditional and **modern drilling methods**, and explore the **role of geophysical surveys**, **hydrogeological mapping**, and other scientific tools in determining water locations.

2.2.1. Traditional Borehole Drilling Methods

a. Manual Drilling (Hand Drilling)

In many rural areas, particularly in developing countries, **manual drilling** has been a traditional method for creating boreholes. This technique involves **digging or drilling by hand** using simple tools such as augers, spades, and shovels. While it is a low-cost and accessible method, it has significant limitations in terms of depth, efficiency, and precision.

- **Process**: A **manual drill** (often a simple auger or handheld rotary drill) is used to create a hole in the ground, which is then progressively deepened by either rotating or lifting and dropping the tool.

- **Depth**: This method is typically used for relatively shallow wells (less than 30–50 metres), often reaching groundwater in **unconfined aquifers**.

- **Labour-Intensive**: Manual drilling is extremely labour-intensive and can take a long time, making it impractical for deeper wells.

- **Limitations**: The depth achievable with manual drilling is limited, and this method is less effective for areas with hard rock formations. Furthermore, locating a suitable water-bearing zone using this technique relies heavily on **local knowledge** (such as divining), as there are few scientific tools to guide the drilling process.

b. Traditional Mechanical Drilling

Traditional mechanical drilling methods include the use of **manual rigs** or simple machines that use **rotary** or **percussion** techniques. These drills can penetrate deeper than manual methods, reaching groundwater stored in deeper aquifers.

- **Percussion Drills**: These rigs rely on a **hammering** action to break through rock layers and create a well. The action of the hammer helps dislodge soil and rock, making it easier to advance the borehole.

- **Rotary Drills**: In this method, a rotating bit is used to cut through layers of soil and rock. Water or drilling fluids are pumped down to **lubricate the bit**, remove debris, and keep the borehole clear.

- **Depth**: Traditional mechanical drilling can reach depths of **50–200 metres**, depending on the terrain and type of equipment used. This technique is often used in **rural areas** for access to groundwater in **confined aquifers**.

- **Limitations**: Although it is more effective than manual drilling, traditional mechanical drilling still has limitations in terms of depth, efficiency, and precision. The lack of **scientific guidance** in locating the right spot can lead to inefficient drilling and failed wells.

Fig. 2 Tube Well Drilling Method

2.2.2. Modern Borehole Drilling Methods

Modern borehole drilling methods have significantly improved over traditional practices in terms of efficiency, depth, precision, and sustainability. These methods combine advanced technologies, geophysical surveys and engineering tools to ensure successful drilling and sustainable groundwater extraction.

a. Rotary Drilling with Advanced Technology

Modern drilling rigs are typically **rotary rigs** that use **mechanised drills** with hydraulic or electric power. These rigs are capable of drilling through much harder geological formations, including rock layers, which were previously difficult to penetrate.

- **Mud Rotary Drilling**: This method uses a drilling fluid (or "mud") to **lubricate** the drill bit, **cool** it, and **clear debris** from the borehole. The fluid also helps maintain the integrity of the hole, especially in loose or unstable soil.

- **Air Rotary Drilling**: This method uses compressed air to lift debris out of the borehole while the drill bit rotates. It's often used for shallower wells and is particularly effective in **soft soil** or **sandstone** formations.

- **Directional Drilling**: A more advanced technique that allows drillers to **steer** the borehole along a specific path to reach a target water-bearing zone. This method is useful when a vertical borehole is not feasible or when drilling multiple wells from a single location.

b. Drilling Depth and Precision

Modern drilling rigs are capable of reaching **greater depths** (up to 1,000 metres or more) and can create **larger-diameter** boreholes. This makes them suitable for **deep aquifers** or areas where **groundwater resources are deeper** or more challenging to access. **Geotechnical surveys** and **hydrological data** help pinpoint the most promising locations for drilling, reducing the risk of failed wells.

c. Advanced Drilling Fluids and Techniques

- **Polymer-based Drilling Fluids**: These are used for deeper drilling projects. The use of specialised**polymers** in drilling fluids improves efficiency and reduces the environmental impact, as they help with better filtration and cleaning.

Hydraulic Fracturing: In some cases, if groundwater is trapped in low-permeability rock, **hydraulic fracturing** (cracking) may be used to enhance water flow into the well.

1. Technology and Equipment

Comparison between these two methods

Aspect	Traditional Methods	Modern Methods
Tools	Manual tools (shovels, pickaxes, bailers, etc.)	Advanced rigs (rotary, DTH, air drills)
Machinery	Minimal or no mechanization	Highly mechanized, powered by engines
Precision	Limited, often dependent on skill	High precision with computer-guided systems
Adaptability	Suitable for soft soils or shallow depths	Can handle varying terrains, including hard rock

2. Depth and Yield

Aspect	Traditional Methods	Modern Methods
Depth Capacity	Limited to 20–50 meters (65–165 feet)	Up to 3,000 meters (10,000 feet)
Aquifer Access	Shallow aquifers only	Can tap into deep aquifers
Water Yield	Generally lower and unreliable	High, with consistent flow rates

3. Speed and Efficiency

Aspect	Traditional Methods	Modern Methods
Drilling Speed	Slow and labor-intensive	Fast, with some methods drilling 500 meters per day
Time Taken	Days to weeks for shallow bores	Hours to days for deep bores
Workforce	Requires a larger manual workforce	Operated by a small, skilled team

4. Cost

Aspect	Traditional Methods	Modern Methods
Initial Cost	Low, as tools are inexpensive	High, due to advanced equipment and machinery
Operational Cost	Low, uses manual labor	High, involves fuel, maintenance, and skilled labor
Long-term Costs	High due to potential inefficiencies and maintenance	Lower, as modern bores are more durable and efficient

5. Suitability and Accessibility

Aspect	Traditional Methods	Modern Methods
Terrain	Effective in soft soils or shallow formations	Suitable for all types of terrains, including hard rock
Infrastructure Needs	Minimal, used in remote areas	Requires access for rigs and machinery
Applications	Small-scale domestic or agricultural purposes	Large-scale irrigation, industries, and urban water supply

6. Environmental Impact

Aspect	Traditional Methods	Modern Methods
Environmental Disruption	Minimal disruption due to manual digging	Can cause more disturbance due to heavy machinery
Groundwater Contamination Risk	Higher, due to improper sealing	Lower, with proper casing and sealing
Aquifer Sustainability	Shallow wells may overdraw small aquifers	Can overexploit deep aquifers if unregulated

7. Maintenance and Lifespan

Aspect	Traditional Methods	Modern Methods
Lifespan	Shorter, prone to collapse	Longer, with durable casings and structure
Maintenance	Frequent manual cleaning needed	Minimal with modern pumps and protection systems

8. Examples of Use

Scenario	Traditional Methods	Modern Methods
Rural Areas	More common where advanced technology is unavailable	Increasingly used, but requires infrastructure
Urban Areas	Rare, impractical for deep aquifers	Standard for municipal and industrial needs
Emergency Situations	Quick and feasible for temporary shallow wells	Used for permanent and high-demand solutions

2.2.3. Role of Geophysical Surveys and Hydrogeological Mapping

Modern drilling methods are greatly enhanced by the use of **scientific tools** and **hydrological surveys** to identify and assess groundwater resources before drilling. These tools help guide the selection of drilling sites, reducing the risks of unsuccessful wells and ensuring more efficient and sustainable groundwater extraction.

a. Geophysical Surveys

Geophysical surveys use **non-invasive techniques** to map subsurface geology and identify water-bearing formations. By analysing the physical properties of the Earth's surface, geophysicists can assess the presence and quality of groundwater and identify areas with the best potential for successful boreholes.

- **Electrical Resistivity Tomography (ERT)**: ERT is used to measure the **resistivity** of soil and rock formations. Water-bearing zones typically have lower resistivity than dry rock or clay. ERT can help map **aquifer boundaries** and provide insights into the **depth** and **extent** of groundwater resources.

- **Seismic Surveys**: Seismic surveys involve sending sound waves through the Earth's surface and measuring their reflections. These surveys help identify **rock layers**, faults, and other geological structures that may affect groundwater flow.

- **Ground-Penetrating Radar (GPR)**: GPR uses high-frequency radar waves to detect subsurface features. It's particularly useful in locating **shallow aquifers** or **fractures** in rock that may contain water.

- **Magnetic and Gravity Surveys**: These surveys can help identify **structures** like faults, fractures, or aquifer boundaries that affect the flow of groundwater.

b. Hydrogeological Mapping

Hydrogeological mapping involves the study of the geology, hydrology, and water quality of a region to understand groundwater resources and their sustainability.

- **Mapping Aquifer Types**: Different types of aquifers (unconfined, confined, semi-confined) require different drilling approaches. Hydrogeological maps help determine which areas are best suited for particular types of bore wells.

- **Recharge and Discharge Zones**: Hydrogeological maps can identify regions where aquifers are recharged (through infiltration from rain or rivers) and areas of discharge (where groundwater seeps out naturally, such as in springs or wetlands). This helps identify **sustainable drilling locations**.

- **Water Quality Assessment**: Hydrogeological surveys also evaluate the **water quality** of potential aquifers, identifying issues like **salinity**, **contamination**, or **mineralisation** that could affect the usability of the water.

c. Integration of Data

By combining **geophysical data**, **hydrogeological mapping**, and **environmental considerations**, drilling engineers can make more informed decisions about where to drill and how deep to go. These data-driven decisions help avoid unnecessary drilling costs and improve the success rates of boreholes.

2.2.4. Advantages of Modern Methods over Traditional Techniques

- **Precision and Efficiency**: Modern geophysical surveys and hydrogeological mapping provide highly accurate data on groundwater availability, ensuring that borewells are drilled in the most promising locations. This reduces the risk of **dry wells** and maximises the likelihood of successful drilling.

- **Deeper and More Reliable Water Sources**: Modern drilling techniques allow access to **deeper aquifers**, providing access to larger volumes of water that are less likely to be affected by seasonal variations or surface contamination.

- **Sustainability**: Using **scientific tools** to assess the **sustainability** of groundwater resources before drilling ensures that water extraction does not exceed the natural **recharge rate**, reducing the risk of **over-extraction** and **aquifer depletion**.

- **Cost-Effectiveness**: Although modern drilling methods are more expensive upfront, they often prove more cost-effective in the long run because they reduce the risk of **failed wells** and **wasted resources**.

2.3. The Science of Groundwater and Water Resource Planning: The Gap in Traditional Engineering

Water resource management, particularly groundwater, is a critical issue in rural and developing areas where access to clean water is often limited or unreliable. Traditional engineering methods, while effective in many contexts, may not always be the most suitable or successful for addressing water challenges in such regions. This is due to a variety of factors, including a lack of access to advanced technology, challenging terrain, and high costs. Let's explore how these factors impact groundwater and water resource planning.

2.3.1. Lack of Access to Advanced Technology

- **Limited Access to Remote Sensing and GIS**: In developed countries, advanced tools such as Geographic Information Systems (GIS), satellite imagery, and remote sensing are used to map water resources and predict groundwater availability and movement. These technologies are expensive and often unavailable in rural or developing areas. Without these tools, groundwater planning becomes less accurate and more reliant on traditional methods that may be imprecise or outdated.

- **Lack of Groundwater Modelling Tools**: Advanced hydrological modelling software (e.g., MODFLOW, hydrogeological models) is essential for understanding the flow and storage of groundwater, predicting aquifer recharge, and planning long-term water management strategies. These models often require high-quality data, significant computational resources, and skilled personnel, all of which are often lacking in rural or developing areas.

- **Limited Monitoring Capabilities**: Traditional groundwater management often relies on physical well measurements and rudimentary data collection. Advanced techniques, such as continuous water quality and quantity monitoring, are less common in rural areas due to both technological and financial constraints. Without real-time data, decision-making for sustainable groundwater use becomes more speculative and reactive, rather than proactive.

2.3.2. Terrain and Environmental Challenges

- **Geological Variability**: Groundwater availability and quality are highly dependent on local geological conditions, which can vary significantly across rural and developing regions. In mountainous, arid, or remote areas, the geology can be complex, making traditional water planning methods less effective. For example, fractured rock aquifers or karst systems may require specialised knowledge and approaches that traditional engineering often doesn't address.

- **Access to Water Sources**: In many rural or developing areas, groundwater resources are located in areas that are difficult to access due to rugged terrain, remote locations, or lack of infrastructure (roads, electricity, etc.). In such cases, traditional engineering approaches—such as drilling deep wells— may be prohibitively expensive or physically challenging. Terrain features such as deserts, mountains, or floodplains can complicate the planning and development of water resources in ways that conventional engineering does not fully accommodate.

- **Water Quality Issues**: Groundwater in rural or developing areas may be contaminated by natural or human activities (e.g., heavy metals, nitrates, saltwater intrusion, pathogens). Traditional water resource engineering methods often fail to account for local water quality issues unless specialised interventions like water purification systems, filtration, or desalination are integrated into the infrastructure, which are often costly and technically demanding.

2.3.3. Cost and Resource Constraints

- **High Initial Costs**: Traditional engineering methods for groundwater extraction, such as drilling deep boreholes, installing pumping systems, and setting up infrastructure for water distribution, can be very costly. In rural or developing regions where financial resources are limited, these upfront costs can be prohibitively high, even if they are necessary for long-term water security.

- **Sustainability and Long-Term Costs**: Groundwater extraction in rural or developing areas is often done without proper long-term planning, leading to over-extraction or depletion of aquifers. In areas where traditional methods like hand-dug wells or shallow boreholes are used, there may be insufficient consideration of the aquifer's recharge rate, leading to unsustainable usage. The lack of advanced modelling or monitoring tools makes it difficult to accurately predict future water availability, exacerbating the cost problem when over-extraction leads to drying wells or saltwater intrusion.

- **Maintenance and Operational Costs**: Groundwater extraction and distribution systems require continuous maintenance. In rural and developing areas, where there may be a lack of skilled labour or technical resources, maintaining these systems becomes a significant challenge. Traditional engineering often relies on centralised, complex infrastructure that may not be well-suited to local conditions or capacity. This can lead to system failures or inefficiencies, increasing the overall cost of water management.

- **High Energy Requirements**: Groundwater extraction systems often rely on electricity or diesel to power pumps. In rural areas with unreliable power grids, or no grid at all, energy costs can be an additional barrier. Without access to renewable energy sources (e.g., solar-powered pumps), the cost of energy for groundwater extraction may exceed what local communities can afford.

2.3.4. Social, Cultural, and Institutional Barriers

- **Community Engagement and Ownership**: Traditional engineering methods tend to focus on large-scale, top-down solutions that may not be appropriate for rural or developing communities. This can lead to a disconnect between the community's needs and the infrastructure that is built. Without active community engagement and input, water management strategies can fail, and the long-term sustainability of water resources is compromised. Local knowledge, which can be crucial in identifying sources and predicting seasonal variations in water availability, may be undervalued.

- **Inadequate Governance and Regulation**: Water resource management in rural or developing areas may be hampered by weak governance, lack of proper regulation, and insufficient local capacity to manage water sustainably. This can lead to over-extraction, contamination, and inefficient use of groundwater. Traditional engineering practices may focus on the construction of infrastructure without addressing these underlying governance issues, which are critical for long-term success.

2.4. Key Takeaways

- **Traditional Methods** are best suited for small-scale, shallow water needs in remote areas with limited resources. They are cost-effective but less efficient and reliable.

- **Modern Methods** offer better efficiency, depth, and water yield, making them ideal for large-scale and long-term applications. However, they are expensive and require skilled operators.

While modern methods have largely replaced traditional ones, a combination of both may be practical in certain regions, especially when complemented with sustainable water management practices.

2.5. Conclusion

Groundwater is a crucial resource for many regions, particularly in rural and resource-poor areas. Understanding the science behind **aquifers, groundwater flow**, and **boreholes** is essential for managing water resources sustainably. While traditional methods of locating water, such as **dowsing** or **divining**, continue to be used in many parts of the world, they have limitations compared to modern **hydrological** and **geophysical** techniques. Modern methods offer more accurate, detailed, and scientifically reliable assessments of water sources, enabling more effective and sustainable water management practices.

Modern borehole drilling practices have significantly advanced over traditional methods, thanks to the integration of **scientific tools**, **geophysical surveys**, and **hydrogeological mapping**. These tools help identify the best locations for drilling, assess the depth and quality of water, and ensure that extraction is sustainable. While traditional methods like manual and mechanical drilling are still used in some areas, modern engineering practices offer more precise, efficient, and environmentally responsible solutions for accessing groundwater resources. As demand for water increases globally, the application of these modern techniques will play a crucial role in ensuring that water resources are managed effectively and sustainably.

◆ ◆ ◆

Chapter 3

Groundwater Divining Meets Engineering: How They Complement Each Other

3.1. The Need for Complementary Approaches:

- Why engineers and water professionals are turning towards complementary techniques to improve the accuracy and success of bore well drilling.

3.2. Case Study: Successful Integration:

- Highlight a project or region where divining and modern engineering methods worked handinhand to locate water sources effectively.

- Discuss the process: How divining was used alongside engineering surveys, and the impact it had on project outcomes.

3.3. The Role of Diviners in Water Resource Planning:

- How professional diviners can work with engineers to increase borewell success rates.

- The importance of local knowledge and the intuition of experienced diviners.

◆ ◆ ◆

Groundwater divining, also known as dowsing, is an ancient practice where individuals claim to locate underground water sources using tools like a dowsing rod or pendulum. While divining has been widely dismissed by many in the scientific community due to the lack of empirical evidence supporting its effectiveness, there is a growing interest in the idea of combining traditional practices like divining with modern engineering techniques to improve groundwater exploration and borehole drilling success.

3.1. The need for complementary approaches

Arises from the limitations of both groundwater divining and traditional engineering methods, especially in rural or developing regions. By integrating these approaches, engineers and water professionals aim to enhance accuracy, reduce costs, and improve the success rates of borehole drilling.

3.1.1. Challenges of Traditional Borehole Drilling

Traditional engineering methods for groundwater exploration typically rely on a combination of geological surveys, hydrological studies, and advanced technologies (e.g. geophysical surveys, satellite imagery, or drilling reports). While these methods can be effective, they often face challenges in rural or remote locations where:

- **Limited Data Availability**: In areas where detailed geological or hydrological data are scarce, traditional methods can become less reliable. Drilling is often expensive, and without accurate data, the risk of unsuccessful borehole drilling increases. Engineers may have to drill multiple test wells at significant cost before locating an adequate water source.

- **High Costs and Time Consumption**: Comprehensive geophysical surveys or drilling multiple exploratory wells can be time-consuming and costly, especially in remote areas with limited infrastructure. Even with modern techniques, there is still a level of uncertainty involved in predicting the best locations for drilling.

- **Geological Variability**: Groundwater availability and the specific characteristics of aquifers (such as depth, flow rate, and quality) can vary dramatically over short distances. Traditional methods might not always capture this variation accurately, leading to difficulties in planning and decision-making.

3.1.2. The Role of Groundwater Divining in Borehole Drilling

Groundwater divining is based on the belief that certain individuals have the ability to detect underground water through heightened sensitivity to changes in the Earth's electromagnetic field or through intuitive methods. Although it is widely regarded as pseudoscience by many, divining has been practised for centuries, and some diviners claim to have success in locating water sources in difficult conditions where modern techniques struggle.

Groundwater divining is seen as having potential complementary benefits in borehole drilling:

- **Identifying Potential Locations**: Divining can help identify areas where engineers might have missed possible water sources, especially in remote or hard-to-access regions. Some practitioners of dowsing claim that they can locate underground fractures, fault lines, or aquifer pathways that would otherwise be difficult to detect using traditional methods.

- **Reducing the Need for Extensive Surveys**: In areas where geological data is sparse, diviners may be able to provide valuable insight into where to drill, thus reducing the number of exploratory boreholes required. This can save significant time and cost in the early stages of water resource development.

- **Locating Shallow Water Sources**: Dowsing is often useful in locating shallow, small aquifers or groundwater pockets that may not be easily detectable through conventional engineering methods, especially in areas with limited access to advanced technology. This can help engineers pinpoint potential shallow wells for water supply in rural or arid regions.

3.1.3. Complementary Techniques – How They Work Together

Engineers and water professionals have started to embrace the idea of combining the intuitive, observational practices of groundwater divining with the rigorous, scientifically grounded approaches of modern engineering. Here's how these approaches can complement each other:

3.1.3.1. Initial Location Identification

- **Divining**: A dowser is often consulted before formal surveys are conducted to identify potential sites where groundwater might be present. While the exact location and depth of water may not be guaranteed, diviners can narrow down areas worth investigating, potentially saving time and resources.

- **Engineering**: After the diviner identifies a potential site, engineers perform more detailed assessments, using geophysical surveys or other methods to verify the presence of water and gather data on the aquifer's characteristics (depth, flow rate, and quality).

3.1.3.2. Verification and Data Collection

- **Engineering**: With an identified location, engineers can employ modern techniques such as electrical resistivity tomography (ERT) or ground-penetrating radar (GPR) to assess soil properties, determine groundwater depth, and estimate flow patterns.

- **Divining**: In areas where geophysical methods are challenging (e.g., uneven terrain, deep aquifers), diviners may offer additional guidance on the specific drilling points or areas where the water source is likely to be concentrated.

3.1.2.3. Reducing Costs and Increasing Efficiency

- **Engineering**: Borewell drilling can be a costly and time-intensive process, especially when unsuccessful attempts are made. Divining can help reduce the risk of drilling in unpromising locations, thereby decreasing the number of unsuccessful borewells and reducing overall costs.

- **Divining**: If a diviner can guide engineers to more promising locations, fewer exploratory wells will be needed, and drilling operations can be more efficient.

3.1.3.4. Conflict Resolution and Community Trust

- **Divining**: In many rural and developing areas, traditional practices and beliefs are deeply rooted in the local culture. Groundwater divining may have strong cultural significance, and local communities may have more trust in a local dowser than in external engineers. By integrating divining into the water exploration process, engineers can build trust and foster collaboration with the community.

- **Engineering**: Engineers can respect the local knowledge and traditions by recognising the value of divining, while also ensuring that scientifically sound and sustainable practices are followed in the actual drilling process.

3.1.3.5. Local Adaptation and Customisation

- **Divining**: In some regions, divining may be more effective in locating certain types of water sources that are difficult to detect using traditional techniques. In arid regions, for example, diviners may be skilled at identifying groundwater sources that are not immediately obvious from geological surveys.

- **Engineering**: Combining divining with engineering allows for customised approaches based on local knowledge and environmental conditions, leading to more appropriate and tailored solutions for water resource planning.

3.1.3.6. Caveats and Limitations of Combining Methods

While the combination of divining and engineering can be complementary, there are a few considerations to keep in mind:

- **Scientific Validation**: Dowsing lacks scientific validation, and its results are often subjective. While it may be helpful in some instances, engineers need to ensure that dowsing is used as a supplementary tool, not a replacement for rigorous scientific methods.

- **Local Expertise**: Engineers and water professionals should ensure that they work with dowsers who have a deep understanding of local geology and hydrology, as dowsing methods vary widely by region and practitioner skill.

- **Regulatory and Ethical Issues**: Some regions or institutions may not allow divining to be incorporated into professional water planning. Engineers must navigate any legal or regulatory frameworks that could impact the acceptance of complementary approaches.

3.2. Case Study 2: Successful Integration of Groundwater Divining and Modern Engineering in Rural India

One notable example of successfully integrating groundwater divining with modern engineering methods is a water supply project in the **Marathwada region of Maharashtra**, India. This region, known for its arid climate and frequent droughts, has faced significant challenges in securing reliable water sources. The project aimed to provide sustainable water access to rural communities by combining traditional groundwater divining with scientific, engineering-based techniques.

3.2.1. Background: The Need for Water

The Marathwada region, which includes parts of Jalna, Beed, and Aurangabad districts, has faced recurring water scarcity issues. Due to the region's dry climate, groundwater is the primary source of drinking and irrigation water. However, the region's complex geology and frequent water shortages made identifying reliable and sufficient water sources difficult. Traditional engineering methods like hydrogeological surveys and geophysical assessments often yielded mixed results, as many local aquifers were shallow, poorly mapped, or had unpredictable recharge rates.

In this context, local communities had long relied on **traditional groundwater divining** as a method to locate underground water sources. While divining was deeply rooted in local practice and culture, its success rates and reliability were questioned by many engineers, leading to scepticism about its integration into formal water planning.

3.2.2. Project Design and Approach

A rural water supply project in the Marathwada region was initiated by the **Maharashtra Groundwater Department** in collaboration with an NGO and local communities. The project aimed to locate water sources for sustainable boreholes and address the growing need for drinking and irrigation water in 20+ villages. To maximise the success of the project, the team decided to **combine modern engineering surveys with traditional groundwater dowsing**. This integrated approach was designed to both increase drilling success rates and reduce the cost and time involved in water resource exploration.

Step 1: Groundwater Divining

The first phase of the project involved **local groundwater diviners** who were known for their skill in locating underground water sources. These diviners used traditional tools, such as **dowsing rods, pendulums**, and **water-witching sticks**, to identify potential locations for drilling. Based on their local knowledge and experience, they marked out areas that they believed would yield water.

- **Community Trust**: Local communities, who had long relied on these diviners for locating water in the past, readily accepted this approach. The diviners worked closely with the engineers and mapped out several potential locations for borewells. The role of the diviners was crucial not just in identifying sites, but also in **building trust within the community**.

Step 2: Modern Engineering Surveys

Once the diviners identified promising sites, modern engineers and hydrologists conducted **geophysical surveys** to verify the presence of groundwater and assess the quality, depth, and flow rates of potential water sources. Methods like **electrical resistivity tomography (ERT)**, **seismic surveys**, and **magnetic resonance sounding (MRS)** were used to gather detailed data on soil composition, aquifer characteristics, and the depth of groundwater.

- **Verification and Data Refinement**: Engineers cross-checked the divining results using scientific methods, confirming that many of the diviners' suggested locations aligned with areas where groundwater was present. In some cases, the diviners had pinpointed water sources that were otherwise hard to detect through traditional surveys due to the complexity of local geology (such as fractured rock or hard-to-map aquifers).

Step 3: Drilling and Borehole Installation

After identifying viable locations, the project team proceeded with drilling boreholes. The **drilling engineers** used the data gathered from the divining and geophysical surveys to select drilling depths and optimise drilling methods. In many cases, the boreholes hit reliable water sources on the first attempt, avoiding the need for additional exploratory drilling, which can be costly and time-consuming.

Step 4: Water Quality Testing and Distribution

Once water was located, the boreholes were tested for quality, ensuring they met drinking water standards. In some cases, additional filtration or purification systems were added to address concerns such as high salinity or contamination from nearby agricultural activities. After confirming the water quality, the team set up **water distribution systems** for the villages, ensuring a sustainable supply of water for drinking and irrigation.

3.2.3. Impact on Project Outcomes

The integration of divining and engineering methods resulted in several positive outcomes:

1. **Increased Drilling Success**: By combining divining with modern engineering surveys, the project achieved a **higher success rate** in locating water sources on the first attempt. Divining helped identify promising locations that engineering surveys alone might have missed due to the challenging terrain and complex geology of the region.

2. **Cost and Time Savings**: The use of divining helped reduce the need for extensive and expensive exploratory drilling. Instead of drilling multiple boreholes based on uncertain geological data, engineers could narrow down potential sites based on the divining results, saving time and resources.

3. **Community Empowerment and Trust**: The project demonstrated the value of incorporating local knowledge and traditions into modern water management efforts. By working with local diviners, the project team gained the trust and support of the communities. This was especially important in a region where people had deep-rooted faith in divining and where scepticism about external engineers and their methods was common.

4. **Sustainable Water Sources**: The project succeeded in providing sustainable and reliable water sources to over 20 villages, improving access to clean drinking water and supporting agricultural activities. The integration of divining also helped identify sources that were capable of sustaining long-term water needs, such as deeper or more resilient aquifers.

5. **Capacity Building**: The project also contributed to **capacity building** within the local community. Local engineers and technicians were trained alongside diviners, learning to appreciate the value of traditional knowledge while reinforcing the importance of scientific validation and monitoring.

3.3. The Role of Diviners in Water Resource Planning: How They Can Work with Engineers to Increase Borehole Success Rates

In regions with scarce or unreliable access to groundwater, traditional practices like groundwater divining (or dowsing) continue to play an important role in the search for water. While often regarded with scepticism in the scientific community, diviners have valuable local knowledge and intuitive skills that, when paired with modern engineering and hydrological methods, can significantly improve the success rates of borehole drilling and water resource planning.

Professional diviners, or water witches, possess deep intuition and experience honed over years or even generations of practice. By understanding local geologies, water flow patterns, and cultural landscapes, they can help engineers identify viable locations for drilling. When combined with engineering expertise, divining can enhance water exploration efforts, especially in complex or poorly mapped terrains.

3.3.1. How Professional Diviners Can Work with Engineers to Increase Borehole Success Rates

3.3.1.1. Early Identification of Potential Water Sources

Professional diviners are often the first step in the process of identifying potential locations for groundwater drilling. Using tools such as dowsing rods, pendulums, or even their own intuition, diviners can indicate areas where underground water might be present. While these methods are not scientifically quantifiable, experienced diviners often have a **heightened sensitivity to subtle environmental factors**, such as changes in soil composition, electrical fields, and other natural indicators that may signal water below the surface.

- **Initial Site Selection**: Diviners can help engineers **narrow down a broad search area** by identifying likely spots where the aquifers may be located. This can be especially valuable in regions where geological surveys are sparse or inconclusive, saving time and reducing costs in the initial stages of water exploration.

- **Identifying Shallow Aquifers or Hidden Water Channels**: In rural and developing areas, **shallow aquifers** or **fractured rock formations** may not be easily detected using conventional geophysical methods. Diviners may be able to locate these sources based on intuitive cues or knowledge of historical water availability in the region, offering valuable leads to engineers.

3.3.1.2. Validation and Verification with Modern Techniques

Once diviners have identified potential water sources, the next step is for engineers to **validate and verify these locations using modern scientific techniques** such as geological surveys, geophysical methods (e.g. electrical resistivity tomography, seismic surveys), or satellite imagery.

- **Complementary Approach**: Divining does not replace scientific methods but can be used as a **complementary tool**. While diviners locate potential areas, engineers can use geophysical surveys to confirm the depth of the water, its flow rate, and quality, ensuring that the groundwater source is suitable for sustainable extraction.

- **Reducing Uncertainty**: Divining can help reduce the **uncertainty of drilling**. If the diviner points to a promising spot, engineers can adjust their drilling plans accordingly, optimising resources and effort to focus on more likely locations, thus avoiding the risks of drilling dry wells.

3.3.1.3. Reducing Costs and Time

Borehole drilling can be a costly and time-consuming process, particularly if multiple exploratory wells are required before a reliable water source is located. By integrating dowsing into the planning process, engineers can focus their drilling efforts on the **most promising sites**, reducing the need for costly and time-consuming exploratory drilling and increasing the overall efficiency of water sourcing.

- **Fewer Unsuccessful Boreholes**: Divining helps reduce the chances of **drilling dry boreholes** or poorly positioned boreholes by providing an intuitive starting point. Diviners' knowledge of local water sources may help engineers avoid areas where water is difficult to extract, such as places with hard, impermeable rock layers or where aquifers have dried up.

- **Faster Deployment**: Combining dowsing with engineering surveys can expedite the process of drilling boreholes. Dowsers can quickly identify potential sites, allowing engineers to perform verification and begin drilling sooner than if they relied solely on conventional geophysical surveys, which can take more time to conduct.

3.3.2. The Importance of Local Knowledge and the Intuition of Experienced Diviners

3.3.2.1. *Local Knowledge and Experience*

Professional diviners typically grow up in the areas where they practice their craft and develop a deep understanding of local landscapes, climate conditions, and historical water sources. This **local knowledge** often allows them to identify patterns and signs that may not be immediately apparent to outsiders or modern engineers.

- **Intuitive Knowledge**: Experienced diviners often rely on **intuition** developed through years of practice and local knowledge. They may recognise subtle geological signs, such as differences in soil texture, vegetation patterns, or natural springs, that indicate the presence of underground water. In some cases, diviners might also consider environmental factors like rainfall patterns, seasonal changes, or the behaviour of local wildlife in relation to water availability.

- **Cultural Significance**: Diviners often play an important role in their communities, where their skills are passed down through generations. As a result, their **relationship with the land** and their understanding of its water systems are often finely tuned. In many rural or developing areas, diviners are trusted by the local population, which can significantly improve cooperation with water resource projects and help overcome cultural barriers.

3.3.2.2. *Intuitive Decision-Making in Challenging Terrains*

In many rural and remote regions, the geology can be complex or poorly understood. This can make it difficult for engineers to reliably predict where water might be located, especially in areas where sophisticated geological surveys are unavailable or infeasible. Diviners, however, may have a unique ability to navigate these challenges due to their **intuition and familiarity with local hydrology**.

- **Identifying Subtle Water Indicators**: Experienced diviners can often sense water in areas where conventional methods may be ineffective. For example, they might identify underground fractures, fault lines, or water channels in areas where engineers might not have sufficient data. In some places, divining has been reported to help locate **underground rivers** or **deep aquifers** that modern technology has not detected.

- **Navigating Complex Geologies**: In terrains with fractured bedrock, karst landscapes, or areas where groundwater is deeply buried, diviners' ability to sense underground water can prove invaluable. This kind of geological complexity can be hard to map using geophysical methods alone, but diviners often use their experience to intuitively locate areas with high potential for successful drilling.

3.3.2.3. *Building Trust and Community Engagement*

In regions where modern engineering and water management may be viewed with scepticism, especially when brought in from external organisations, involving local diviners in the water resource planning process helps **build trust** with the local community. Diviners are respected members of the community, and their involvement can lead to greater acceptance of engineering projects and encourage local participation.

- **Cultural Sensitivity**: In many rural areas, the acceptance of externally driven projects is often hindered by **mistrust** of modern technology or outside experts. When diviners are involved, it helps bridge the gap between traditional knowledge and modern engineering, fostering cooperation between **scientific experts** and **local communities**.

- **Local Empowerment**: By integrating diviners into water resource planning, the community feels that their traditional knowledge and practices are being respected and utilised in a **practical way**. This empowerment can lead to more sustainable and long-lasting water management solutions.

3.3.3. Case Example: Successful Integration of Diviners and Engineers

In several regions of **rural India** (such as in parts of Maharashtra, Gujarat, and Rajasthan), projects have integrated professional diviners with engineers to improve the success of borewell drilling and groundwater management. For instance, in the **Marathwada region of Maharashtra**, diviners worked alongside hydrologists and engineers to successfully locate water in otherwise dry or hard-to-map areas.

- **Diviners identified potential borewell locations**, especially for shallow aquifers that were not easily detectable with traditional methods.

- **Engineers validated these sites** using geophysical surveys, which helped confirm that the identified areas had water at the required depths and flow rates.

- The collaboration resulted in **higher drilling success rates** and reduced costs, as multiple dry wells were avoided, and the borewells hit the water on the first attempt.

This integration not only ensured the success of the project but also helped maintain **community involvement and trust**, as the role of diviners was valued alongside scientific methods.

3.4. Lessons Learned and Key Takeaways

3.4.1. **Respect for Local Practices**: One of the most important takeaways from the project was the recognition that traditional practices like divining could play a **valuable complementary role** in modern water management. Engineers and water professionals can achieve better outcomes by respecting local knowledge while integrating it with modern scientific techniques.

3.4.2. **Holistic Approach to Water Resource Management**: The project highlighted the importance of **taking a holistic approach** to water resource planning, one that combines the strengths of both traditional and modern methods. While divining alone might not always provide reliable results, when paired with engineering surveys, it can significantly increase the likelihood of successful water discovery.

3.4.3. **Cultural Sensitivity and Community Engagement**: Building trust within local communities is key to the success of water projects, especially in rural or developing regions. By involving diviners—respected members of the community—engineers were able to foster a sense of **collaboration and inclusivity**, which made the project more successful and sustainable.

3.5. Conclusion

The Marathwada groundwater project demonstrates how traditional techniques like groundwater divining can complement modern engineering methods to improve the success of borewell drilling and water resource management. This integration not only enhanced the efficiency of water sourcing but also empowered local communities, ensuring that the project was both technically sound and culturally sensitive. As water scarcity continues to be a pressing issue in many parts of the world, combining traditional knowledge with modern engineering offers a promising pathway to more sustainable and inclusive water management solutions.

3.5.1. A Holistic Approach to Groundwater Exploration

The integration of groundwater divining and traditional engineering techniques is not about replacing one method with another but rather recognising that each approach has its strengths and limitations. In areas where modern engineering methods face constraints due to a lack of data, difficult terrain, or cultural barriers, combining these methods can lead to more successful and sustainable water resource planning. By respecting both traditional knowledge and scientific techniques, engineers and water professionals can increase the accuracy, efficiency, and cost-effectiveness of borehole drilling operations, ultimately improving access to clean, reliable water in rural and developing areas.

3.5.2. A Complementary Approach for Sustainable Water Resource Management

The collaboration between professional diviners and engineers can significantly enhance the success of groundwater exploration and borewell drilling efforts. By respecting and incorporating local knowledge, diviners help identify promising water sources that modern engineering might otherwise overlook. When paired with scientific validation through modern technologies, this complementary approach increases the likelihood of successful water sourcing, **reduces costs**, and fosters **community trust** in water resource projects.

This synergy between traditional knowledge and modern techniques represents a **holistic approach** to sustainable water management that can be particularly effective in rural and developing regions where both local expertise and scientific tools are needed to address complex water challenges.

◆　◆　◆

The Hidden Blueprint: Understanding Earth's Structural Framework

4.1. Engineering Practices in Ten different portals across the Globe: geospatial and geophysical data layers, including lineaments, geological features, and satellite imagery:

- Websites.

- The Descriptions.

- Methods.

4.2. The Bhuvan Portals:

- Steps for Detecting Lineaments.

- Key features, examples and workflows.

- Conclusions.

4.3. General Steps to Detect Lineaments Using Remote Sensing and GIS:

- Map pictures for Bhuvan lineaments.

◆ ◆ ◆

4.1. Engineering Practices in different 10 portals across the globe

To access lineament data (fractures, faults, and linear features) for different regions of the world, you can explore the following reference portals, similar to **Bhuvan**, which provide various geospatial and geophysical data layers, including lineaments, geological features, and satellite imagery:

4.1.1. USGS Earth Explorer

- **Website**: https://earthexplorer.usgs.gov/

- **Description**: The U.S. Geological Survey (USGS) Earth Explorer provides access to a wide range of satellite and aerial imagery, including datasets useful for identifying geological features like lineaments. The platform offers free access to various datasets such as Landsat, MODIS, and others that can be used for geological mapping and analysis of lineaments worldwide.

- **Method**:

  - **Image Selection**: Download relevant satellite imagery (e.g., Landsat, Sentinel-2) using Earth Explorer or Sentinel Hub.

  - **Preprocessing**: Apply corrections (radiometric, atmospheric) to improve image quality.

  - **Enhancement**: Use contrast enhancement techniques like **high-pass filtering** or **edge detection** (Sobel, Canny filters) to highlight linear features in the image.

  - **Manual Detection**: Visualise the imagery and manually trace lineaments.

  - **Automated Detection**: Use software tools like **QGIS** or **ArcGIS** to apply algorithms like **Hough Transform** for line detection.

4.1.2. Google Earth Engine

- **Website:** https://earthengine.google.com/

- **Description**: Google Earth Engine is a powerful platform for processing and analysing geospatial data. You can use satellite imagery, topographic maps, and other data layers to study lineaments, geological features, and natural hazards worldwide. It also offers various analysis tools to identify and map geological lineaments.

- **Method:**

  - **Image Collection**: Use GEE to access global satellite imagery (e.g., Landsat, Sentinel, MODIS).

  - **Data Processing**: Apply algorithms to enhance linear features. For example, use **NDVI (Normalised Difference Vegetation Index)** or **NDBI (Normalised Difference Built-up Index)** to highlight geological features.

  - **Edge Detection**: Apply **edge detection filters** or **gradient-based techniques** (like Canny) in GEE's JavaScript or Python API to identify lineaments.

  - **Topographic Analysis**: Use digital elevation model (DEM) data available in GEE to detect linear features in elevation, which may correspond to faults or fractures.

4.1.3. Sentinel Hub

- **Website:** https://www.sentinel-hub.com/

- **Description**: Sentinel Hub provides access to satellite data from the European Space Agency's (ESA) Sentinel satellites. The platform offers advanced tools for visualising and analysing multispectral imagery, which can be used to detect geological features such as lineaments. The service includes tools like EO Browser, which can be used for identifying lineaments in satellite imagery.

- **Method**:

  - **Radar Interferometry**: Sentinel-1 offers **InSAR** (Interferometric Synthetic Aperture Radar) data, which is useful for detecting **surface displacement** due to tectonic activity.

  - **Lineament Detection**: Using **Differential InSAR** or **Amplitude Change Detection**, one can highlight subsurface faults and fractures that manifest as lineaments on the ground.

  - **Edge Detection**: Apply **edge-detection algorithms** in GIS or remote sensing software to highlight linear geological features.

4.1.4. NASA World Wind

- **Website**: https://worldwind.arc.nasa.gov/

- **Description**: NASA WorldWind is an open-source virtual globe that provides access to various geospatial datasets, including satellite imagery and topographic maps. While it is not focused solely on lineament identification, the available imagery and terrain data can be used for geological analysis across different regions.

- **Method**:

  - **Layer Selection**: Choose topographic or satellite imagery layers (e.g., SRTM, MODIS).

  - **Visualisation**: Use WorldWind's 3D visualisation tools to interact with the topography.

○ **Manual Mapping**: Lineaments are manually identified by visually inspecting terrain anomalies or fault zones, often in conjunction with other layers like geological maps or fault datasets.

4.1.5. SRTM Data (Shuttle Radar Topography Mission)

- **Website**: https://www2.jpl.nasa.gov/srtm/

- **Description**: The SRTM dataset provides global topographic data, including elevation and terrain information, which can be used for geological feature identification, including lineaments. The data can be downloaded and used in GIS software for analysis.

- **Method**:

  ○ **Elevation Data**: Use SRTM's digital elevation models (DEMs) to analyse terrain.

  ○ **Slope Analysis**: Run a **slope analysis** in GIS software (like QGIS or ArcGIS) to highlight areas of significant terrain change, which can indicate faults or fractures.

  ○ **Shaded Relief**: Apply a **hillshade technique** to enhance subtle linear features that may correspond to geological structures like lineaments.

4.1.6. GEOSS (Global Earth Observation System of Systems)

- **Website**: http://www.geoportal.org/

- **Description**: GEOSS is a collaborative initiative that provides access to a wide range of geospatial data. It includes data for various environmental and geological features and can be useful for identifying lineaments and other geological formations. You can search through global datasets and use the portal's search tools to find relevant geospatial data.

4.1.7. OpenTopography

- **Website**: https://opentopography.org/

- **Description**: OpenTopography offers high-resolution topographic data that can be used to identify geological lineaments, faults, and other landscape features. It provides access to LiDAR and DEM (Digital Elevation Model) datasets, which can be analysed to map linear features in different regions around the world.

- **Method**:

  - **LiDAR Data**: Download LiDAR or high-resolution DEM data, which provides detailed topographic information.

  - **Slope & Aspect Analysis**: Perform **slope** and **aspect** analyses in GIS software (QGIS/ArcGIS) to detect linear features based on terrain curvature. Lineaments often correspond to fractures or faults where the landscape shows abrupt changes in slope or direction.

  - **Geomorphological Mapping**: Use LiDAR data to map geomorphological features like scarps, ridges, and valleys, which often coincide with lineaments.

4.1.8. Geospatial Data Analysis Tools (ArcGIS Online)

- **Website**: https://www.arcgis.com/home/index.html

- **Description**: Esri's ArcGIS Online platform offers various geospatial analysis tools that can be used for geological mapping and identifying lineaments. It includes access to various datasets, and you can overlay topographic, geological, and satellite imagery for detailed analysis of linear features.

- **Method**:

  - **Topographic and Satellite Data Integration**: Combine topographic data (e.g., DEMs, SRTM) with satellite imagery in a GIS environment.

- **Lineament Extraction Tools**: Use specific tools like **lineament extraction** or **Hough Transform** for automatic lineament identification. These tools detect linear features based on changes in pixel values (e.g., edges, gradients).

- **Fractal Dimension Analysis**: Apply **fractal dimension** techniques to analyse the complexity of the lineaments and detect fault zones in the data.

4.1.9. Geoscience Australia

- **Website:** https://www.ga.gov.au/

- **Description**: Geoscience Australia offers access to geological, geophysical, and geospatial datasets that include geological lineaments, fault lines, and related features. While it primarily focuses on Australian data, the platform also provides access to global datasets and can be a valuable resource for geological analysis.

- **Method:**

  - **Fault and Lineament Databases**: Use pre-existing geological datasets from the Global Fault Database or Geoscience Australia. These databases contain fault lines and lineament data, often derived from geological surveys, which can be directly overlaid on satellite imagery.

  - **Feature Extraction**: Use GIS tools to extract lineaments or geological features from vector data, and refine the analysis using geospatial filtering techniques.

4.1.10. Global Fault Database (GFD)

- **Website:** https://www.gfdonline.org/

- **Description:** The Global Fault Database provides information on the locations and characteristics of faults around the world.

It can be used to study and map fault lines, which are often associated with lineaments. This resource is more specialised in fault and tectonic data but is valuable for lineament identification.

- **Method**:

  - **Database Query**: Access and query the fault data to retrieve existing lineaments and faults.

  - **Overlay**: Overlay the fault database on satellite imagery or DEMs in GIS software to validate the presence of lineaments.

  - **Field Validation**: In some cases, further field verification (e.g., geological surveys) may be required to confirm the features detected as lineaments.

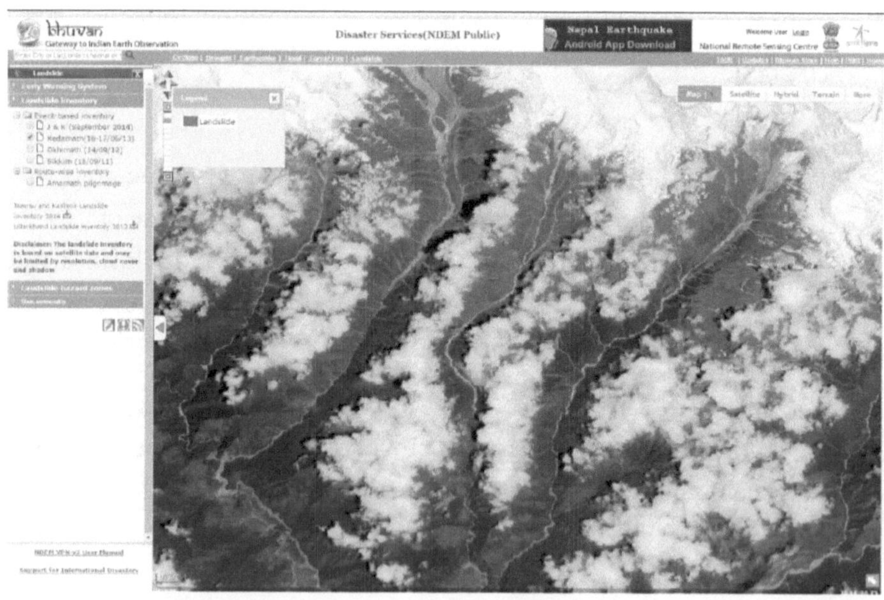

*Fig. 3 Landslide Inventory Map of Uttarakhand 2013 Disaster
(In courtesy of Bhuvan User Handbook)*

4.2. Bhuvan - Bhujal (Groundwater Prospects and Quality Information System) Portal (India) under National Rural Drinking Water Programme

Bhuvan, developed by **ISRO (Indian Space Research Organisation)**, is an online geospatial platform that provides access to satellite imagery, topographical maps, and remote sensing data for various applications, including **lineament detection**. It can be a valuable tool for geologists, environmental scientists, and urban planners in identifying linear geological features such as faults, fractures, and other structural lineaments that influence landforms and hydrology.

4.2.1. Steps for Detecting Lineaments Using Bhuvan Portal

1. **Accessing the Bhuvan Portal**:

   - Visit the Bhuvan website: Bhuvan Portal.

   - Create an account or log in if you already have one to access the full functionality.

2. **Select Appropriate Data Layers**: Bhuvan provides access to multiple satellite datasets that can be used to detect lineaments.

   - **Satellite Imagery**: You can use satellite images such as **Landsat, Cartosat,** or **IRS** data for mapping lineaments.

   - **Topographic Data: Digital Elevation Models (DEMs)** and **terrain data** can help in identifying topographic anomalies or structural features along lineaments.

   - **Soil and Land Use Data:** For identifying regions where linear features correspond to geological formations or specific types of soil.

3. **Choose the Area of Interest:**

   - Use the **zoom-in tool** on the Bhuvan interface to focus on the area where you suspect lineaments may be present, such as fault zones or tectonically active regions.

- **Search by Coordinates**: You can directly input coordinates or select regions based on maps and imagery.

4. **Visual Interpretation of Satellite Imagery**:

   - **Satellite imagery** (e.g. Cartosat, IRS, or Landsat) available on Bhuvan can be used to visually detect lineaments.

     - **Linear Features**: Look for straight or **linear features** such as valleys, river courses, ridges, or fault scarps that may indicate the presence of structural lineaments.

     - **Contrast Enhancement**: Use the available **image enhancement tools** to improve the contrast, which can help highlight subtle linear features in the landscape.

5. **Topographic and Elevation Data**:

   - **DEM Analysis**: If topographic data is available (e.g., DEM), conduct a **slope analysis** or **hillshade analysis**:

     - Areas with **high slopes** or **sudden changes in elevation** (such as cliffs or ridges) can point to tectonic features like faults and fractures that form lineaments.

     - **Linear Valleys or Ridges**: Look for valleys or ridges that align in straight lines, which could correspond to geological fault lines.

   - **Contour Maps**: In topographic data, linear features can appear as straight **contours** or **straight valley axes**, which are indicative of fault-controlled landforms.

6. **Use Image Processing Tools for Lineament Extraction**:

   - **Bhuvan's GIS Tools**: Bhuvan may offer certain tools for lineament extraction in combination with other **GIS** software such as **QGIS** or **ArcGIS**. You can import Bhuvan imagery into these platforms and apply algorithms like **edge**

detection or **Hough transform** to automatically extract lineaments from imagery.

- In **QGIS**, for example, you can use the **lineament extraction** plugin or apply **edge detection filters** to highlight linear features.

- **Canny Edge Detection** or **Sobel Filters** can help emphasise faults and fractures in the imagery, making them easier to trace.

7. **Correlation with Geological Data:**

- **Geological Layers**: You can overlay **geological maps** (if available) with the satellite imagery on Bhuvan to cross-check the linear features with known fault zones or fractures. This helps validate the lineaments identified on satellite images.

  - **Faults and Joints**: Areas of interest where there is a high likelihood of **tectonic activity** (such as mountains, rift valleys, etc.) can often show **alignment of features** that correspond to fault lines or fractures.

8. **Field Validation (if applicable):**

- Once lineaments are detected using Bhuvan, you can plan field visits to the identified regions to validate the presence of faults or fractures. Use **GPS coordinates** to navigate to suspected lineament locations and confirm findings.

Example:

- You observe a straight **ridge** or **linear valley** in a tectonically active region (e.g., the Himalayan foothills) through Bhuvan's topographic or satellite imagery. Upon closer inspection and field validation, you may confirm that this linear feature aligns with a known fault or fracture zone.

9. **Export Data for Further Analysis**:

   - After detecting lineaments using Bhuvan, you can export the imagery or processed data for further analysis in specialised GIS software.

   - **Vector Data Export**: Export the identified lineaments as vector files (e.g. shapefiles) for more detailed mapping and further analysis in other GIS tools.

4.2.2. Key Features of Bhuvan for Lineament Detection

- **High-Resolution Satellite Imagery**: Available from **IRS**, **Cartosat**, and **Landsat** satellites, providing high-quality visual data to detect linear features in the landscape.

- **Topographic Data**: **DEM** and **contour maps** to detect elevation anomalies and structural lineaments like ridges, valleys, and fault zones.

- **Interactive Tools**: Zoom, pan, and measure tools to interact with the satellite imagery and topographic maps.

- **Geospatial Layering**: Ability to overlay **drainage patterns**, **geological maps, landuse data**, and **satellite imagery** for integrated analysis.

4.2.3. Example Workflow Using Bhuvan

1. **Select Region**: Zoom into the region of interest (e.g., a tectonically active area like the Himalayan region or the Deccan Plateau).

2. **Choose Satellite Imagery**: Use available **IRS** or **Cartosat** satellite imagery for the area. Choose the appropriate imagery based on resolution and time period.

3. **Enhance Image**: Apply contrast or enhancement filters to make linear features like faults or fractures more visible.

4. **Overlay DEM**: Add a **DEM layer** to analyse **elevation differences** and identify fault zones that might correspond to linear features.

5. **Examine Drainage Patterns**: Look for **rectilinear drainage** or **displaced stream courses**, which often align with faults or fractures.

6. **Validate Features**: Use **field data** or existing **geological maps** to verify the lineaments identified on Bhuvan.

4.2.4. Advantages of Using Bhuvan for Lineament Detection

1. **Wide Range of Data**: Access to **multiple datasets** like satellite imagery, DEMs, geological data, and more, all integrated into one platform.

2. **Free Access**: Unlike some proprietary mapping platforms, Bhuvan provides **free access** to a variety of satellite imagery and geospatial data, making it accessible for researchers and students.

3. **High-Resolution Imagery**: The availability of high-resolution satellite data (especially **Cartosat** and **IRS imagery**) allows for detailed mapping of linear features and tectonic structures.

4. **GIS Integration**: Bhuvan allows users to download data, making it easy to import into GIS software for more advanced lineament analysis.

5. **Field Navigation**: Bhuvan provides tools to extract **GPS coordinates**, making it easier to navigate to lineament locations in the field for validation.

4.3. General Steps to Detect Lineaments Using Remote Sensing and GIS

1. **Select Data**: Choose appropriate satellite imagery (e.g., Landsat, Sentinel-1, SRTM) or topographic data (e.g., DEM, LiDAR).

2. **Preprocess**: Correct and enhance the data (e.g., atmospheric correction, edge enhancement).

3. **Feature Extraction**: Use edge-detection filters, Hough Transform, or manual interpretation to extract linear features from the imagery.

4. **Topographic Analysis**: Perform slope, aspect, and hillshade analyses to detect linear features in the terrain.

5. **Mapping**: Use GIS software to manually trace or automatically extract lineaments from the processed imagery and data.

6. **Validation**: Compare extracted lineaments with known geological features or fault databases for accuracy.

By using the tools above in combination, you can effectively detect and map lineaments globally. The detection process often involves a mix of automated methods (e.g., edge detection, lineament extraction) and manual interpretation, depending on the quality of the data and the region being studied. These portals offer access to satellite imagery, geospatial data, and analysis tools that can help identify and analyse geological lineaments in various regions of the world. Depending on your needs (e.g., resolution, data format, analytical capabilities), one or more of these platforms may be suitable for your research. By using the tools above in combination, you can effectively detect and map lineaments globally. The detection process often involves a mix of automated methods (e.g., edge detection, lineament extraction) and manual interpretation, depending on the quality of the data and the region being studied.

4.4. Conclusion

The **BhuvanPortal** is an excellent tool for India to detect lineaments due to its access to high-resolution satellite imagery, topographic data, and hydrological maps. By combining visual interpretation with advanced GIS tools and analysis, users can effectively identify and map linear geological features such as faults, fractures, and tectonic boundaries. This makes it a valuable resource for geological studies, disaster management, and environmental assessments.

◆ ◆ ◆

Chapter 5

Visual Pathways: Visual Mapping for Geophysical Lineament Analysis

5.1. Visual Mapping for Geophysical Lineament Analysis.

- 8 Data Maps with Analysis, Examples.

- Limitations of traditional methods of locating water sources.

5.2. Hydro Geo morphological Maps:

- Analysing properties methods.

- Steps for lineament descriptions.

5.3. The Map pictures: Practical examples with pictures.

◆　◆　◆

5.1. Visual Mapping for Geophysical Lineament Analysis

Detecting **lineaments** in the field involves using various types of maps and representations to analyse geological features, landforms, and structural patterns. These maps are typically based on satellite imagery, remote sensing data, and topographic analyses. Below are practical map representations and techniques for detecting lineaments in the field:

5.1.1. Topographic Maps (Slope and Aspect Maps)

- **Use: Topographic maps** provide valuable terrain information that helps in detecting lineaments through **slope** and **aspect** analyses. Lineaments often coincide with faults, fractures, or other geological structures that cause significant changes in elevation and slope.

- **How to Detect Lineaments:**

  - **Slope Map**: A **slope map** highlights areas where there is a steep change in elevation, which could correspond to a fault, ridge, or scarp. Lineaments often manifest as linear features with high or low slope values.

  - **Aspect Map**: An **aspect map** shows the direction of the slope, which can help identify linear patterns that indicate geological features like faults, ridges, and valleys. For example, a linear alignment of east-west slopes may indicate a fault zone.

- **Field Application:**

  - In the field, you can use a **handheld GPS** to navigate to areas with pronounced slope or aspect changes and observe if these correlate with visible linear features (e.g., cliffs or straight valleys).

  - **Geological Mapping**: When combined with geological survey data, these maps help guide field teams to inspect suspected fault zones or fractures in the terrain.

Example

- **Slope Map** of a region shows high slopes along a straight line, indicating a possible fault line. The field team can then inspect this region for fault outcrops or structural deformations.

5.1.2. Satellite Imagery with Enhanced Filters

- **Use**: Remote sensing maps, such as **Landsat** or **Sentinel-2** imagery, can be processed to highlight linear geological features. Enhanced filters such as **edgedetection**, **contrast enhancement**, and **gradient analysis** are used to emphasise linear features that are typically associated with faults and fractures.

- **How to Detect Lineaments**:

  - **Edge Detection**: Use **Canny**, **Sobel**, or **Prewitt filters** to enhance the edges in satellite images, which often correspond to geological lineaments like faults.

  - **Contrast Enhancement**: Applying contrast filters increases the visibility of subtle linear features in the imagery.

  - **Geological Mapping**: A field team can use these enhanced images to navigate to regions of interest where linear features are visible on the ground.

- **Field Application**:

  - Using **Google Earth** or **ArcGIS** mobile applications, field teams can view high-resolution satellite imagery and identify regions where linear features are visible. These can be correlated with observed landforms such as cliffs, straight valleys, or river alignments.

Example

- A **Sentinel-2 image** processed with edge detection reveals a long linear feature, potentially a fault. The field team could examine this region to see if it matches with known geological structures like fault scarps.

5.1.3. Geological Maps

- **Use: Geological maps** provide detailed information about the surface geology of an area, showing the location of rock units, fault lines, fractures, and other geological structures. These maps are critical in identifying and verifying lineaments.

- **How to Detect Lineaments**:

  - **Fault Zones**: Geological maps often display **faults**, **fractures**, and **tectonic boundaries** as lines or zones. These are directly related to lineaments that have influenced the landscape.

  - **Lithological Boundaries**: Changes in rock types (e.g., from granite to shale) often align with lineaments, and geological maps show these boundaries.

- **Field Application**:

  - Field geologists can use **geological maps** to locate known fault zones or fractures and verify the presence of lineaments in the terrain.

  - In the field, the team can look for physical signs of geological structures such as **fault scarps, slickensides** (polished fault surfaces), and other features indicative of faulting or fracturing.

Example

- A geological map indicates a **reverse fault** in an area, and a corresponding field survey identifies a linear valley or cliff that aligns with the fault line. This can be traced further to map the fault in detail.

5.1.4. LiDAR (Light Detection and Ranging) Data

- **Use: LiDAR** provides highly accurate 3D data on terrain elevation and surface structure. This can be crucial in detecting lineaments, especially in regions with subtle or buried geological features.

- **How to Detect Lineaments:**

  - **DEM (Digital Elevation Model):** LiDAR-derived DEMs provide high-resolution elevation data, which is ideal for detecting linear features like faults, fractures, and ridges.

  - **Hillshade:** LiDAR data can be used to create **hillshade maps**, which help in visualising subtle linear features such as fault scarps or fractures.

  - **Slope Analysis:** The steepness of terrain revealed by LiDAR can indicate faults or fractures.

- **Field Application:**

  - LiDAR data can guide the team to areas where there are noticeable elevation differences that may suggest linear features. In the field, they can then confirm these features by looking for fault scarp exposure, displaced strata, or other geological structures.

Example

- A **LiDAR-derived DEM** reveals a subtle linear feature, which the field team investigates to find a fault scarp marking the lineament.

5.1.5. Geophysical Maps (Gravity and Magnetic Data)

- **Use:** Geophysical methods such as gravity and magnetic surveys provide insights into subsurface structures that can indicate lineaments. Faults and fractures typically cause anomalies in gravitational and magnetic fields.

- **How to Detect Lineaments**:
  - **Gravity Anomalies**: Lineaments may be identified by analysing **gravity anomalies,** as faults and fractures can cause variations in the density of underlying rock units.
  - **Magnetic Anomalies**: **Magnetic surveys** detect changes in magnetic susceptibility along linear features. Fault zones or lineaments can cause distinct magnetic anomalies due to changes in rock type or alteration.
- **Field Application**:
  - **Gravity and magnetic maps** can guide field teams to conduct ground surveys in areas with significant anomalies. In the field, lineaments might be observed as surface features or as changes in lithology associated with faulting.

Example

- A **gravity anomaly map** shows a linear anomaly that coincides with a known fault zone. The field team can then investigate this area to look for fault scarps, fractures, or other geological signs that support the lineament hypothesis.

5.1.6. Remote Sensing (SAR/InSAR) Data

- **Use: Synthetic Aperture Radar (SAR)** data and **InSAR** (Interferometric SAR) are particularly useful for detecting surface displacement along lineaments due to fault movement or tectonic activity.
- **How to Detect Lineaments**:
  - **Displacement Detection**: InSAR detects even small **displacements** caused by fault movements, which can help identify active fault zones or fractures.

- ○ **Lineament Extraction**: Automated algorithms process SAR images to extract lineaments based on displacement patterns, often in areas affected by earthquakes or tectonic stress.

- **Field Application**:

  - ○ **InSAR maps** highlight areas of surface displacement, which the field team can visit to examine fault-related features such as fault scarps or offset landforms.

Example

- **InSAR data** shows a displacement pattern that corresponds with a fault line. In the field, the team investigates the area for evidence of tectonic activity and surface rupture.

5.1.7. GIS-Based Lineament Mapping

- **Use: GIS tools** like **ArcGIS** or **QGIS** allow for the integration of different data layers (satellite imagery, DEMs, geological maps) to detect and map lineaments.

- **How to Detect Lineaments:**

  - ○ **Lineament Extraction Tools**: GIS software has automated tools (e.g., **Hough Transform**, **Edge Detection**) to identify linear features in raster data.

  - ○ **Data Overlay**: By overlaying various layers (e.g., geological structures, slope maps, fault data), GIS helps visualise and validate lineaments in the context of the surrounding geological setting.

- **Field Application:**

  - ○ Once lineaments are identified in GIS, field teams use GPS devices to navigate to specific coordinates and conduct detailed geological surveys to confirm the presence of the lineaments.

Example

- A **GIS-generated lineament map** is used in the field to identify possible fault zones, and a geologist surveys the area for visible signs of faulting.

5.1.8. The Global Groundwater Information System (GGIS)

- It can be a valuable resource for accessing hydrogeological maps and groundwater data from various countries around the world. Here's how you can use it:

- **Visit the GGIS Website**: You can access GGIS through the International Groundwater Resources Assessment Centre (IGRAC) website.

- **Search for Data**: The platform allows you to search for groundwater data and information by country, region, or specific parameters.

- **Download Maps**: You can view and download available hydrogeological maps and related data.

- **Interactive Tools**: GGIS provides interactive tools to visualise and analyse groundwater data.

5.2. Hydro Geo Morphological Maps

5.2.1. Ground Water Prospects Maps or HGM-[India] Based Lineament Mapping

In the intricate tapestry of Earth's surface, hidden patterns and alignments silently shape our landscapes, ecosystems, and human activities. These subtle yet powerful features—Ground Water Prospect Maps (GWPMs) called hydrogeomorphic (HGM) maps define systems and lineaments—are the threads that interconnect geology, hydrology, and ecology. While they often elude casual observation, their influence is profound, guiding the flow of water, the formation of soils, and even the distribution of vegetation and wildlife. A user manual for GWPMs is available on the Rajiv Gandhi Drinking Water Mission Projects website.

This chapter unravels the science and significance of HGM mapping and lineament analysis, offering a lens to decode Earth's dynamic processes. From tracing ancient fault lines to understanding the hydrological pathways sculpted by time, we will explore how these tools bridge the gap between raw geology and its practical implications for sustainable resource management, disaster resilience, and landuse planning.

By delving into the methodologies, applications, and case studies, this chapter promises to not only illuminate the hidden structures beneath our feet but also inspire a deeper appreciation of the interconnectedness that defines our natural world.

HGM (High-Grade Map) maps, often used for geographic and topographic purposes, are typically associated with specialised mapping systems for navigation, geological surveys, or other scientific applications. However, you may be referring to specific types of maps or systems like "High-Resolution Digital Elevation Models" (DEM) or even something like "High-Resolution Geospatial Models."

If you're asking about where HGM maps are available, the availability depends on the mapping service or agency involved. Some possible sources for HGM or similar high-resolution maps include:

1. **National Mapping Agencies**: Many countries have national agencies that create high-resolution geospatial and topographic maps, such as:

    - United States: USGS (United States Geological Survey) offers high-resolution DEM data.

    - European countries: Various European countries' national geospatial agencies provide HGM or equivalent high-resolution topographic data (e.g., IGN in France, Ordnance Survey in the UK).

- Canada: Natural Resources Canada (NRCan) provides high-resolution data.

- Australia: Geoscience Australia offers high-resolution geospatial data.

2. **Global Services**: Some global map providers and agencies like NASA, ESA (European Space Agency), and private companies (like Google Earth) also offer high-resolution satellite imagery and elevation models that could be categorised under HGM.

3. **Commercial Providers**: Some private companies, such as DigitalGlobe or Airbus, provide commercial high-resolution mapping and imagery services worldwide.

If you're looking for a specific type of HGM map for a particular country, or if "HGM" refers to something more specific in your context, let me know so I can refine the answer!

Groundwater Prospects Maps or HGM (Hydrogeomorphologic) Maps are a powerful tool in the analysis of lineaments, as they integrate hydrological (water-related) and geomorphological (landform-related) data to provide a comprehensive understanding of landscape features. When analysing lineaments using HGM maps, the approach involves examining how linear features like faults, fractures, and structural alignments influence both surface and subsurface processes (e.g., groundwater flow, and erosion patterns).

Here's how **HGM maps** are used in lineament analysis:

1. Understanding HGM Maps in Context

- **Hydro geomorphological maps** combine information from various datasets, such as:

  - **Topography** (elevation, slope, and aspect maps)
  - **Soil characteristics**

- **Landforms and geomorphological features** (rivers, valleys, fault scarps)

- **Hydrological data** (watersheds, drainage patterns, groundwater flow)

- The maps often represent **drainage networks, flood zones, groundwater recharge areas**, and **geomorphic landforms**, all of which can be influenced or controlled by **lineaments**.

5.2.2. Methods of Lineament Detection Using HGM Maps

A. Analysing Drainage Patterns

- **Method**: Lineaments often influence **drainage patterns** because faults and fractures can act as pathways for water flow, directing streams and rivers along linear features.

- **How to Detect Lineaments**:

  - **Look for linear or rectilinear drainage patterns**: Streams or rivers following straight, linear paths can indicate underlying faults or fractures (tectonic lineaments).

  - **Stream deflection**: Observe areas where watercourses appear to be **deflected or displaced** along straight paths. This can indicate a fault or lineament controlling the flow.

  - **Dendritic vs. Rectangular Patterns**: A **dendritic drainage pattern** typically indicates uniform bedrock, while a **rectangular drainage pattern** may suggest a more structured, fault-controlled terrain, which can be associated with lineaments.

- **Field Application**: A **GIS-based HGM map** showing drainage patterns can be used to locate areas of linear deflection or fault-controlled rivers, which can be further investigated in the field to confirm the presence of geological lineaments.

B. Identifying Landforms and Geomorphology

- **Method**: Geomorphic features such as **valleys**, **scarps**, **ridges**, and **mountain fronts** are often aligned with lineaments, particularly faults or fractures. These landforms can be used to map lineaments.

- **How to Detect Lineaments**:

  - **Fault Scarp Detection**: A prominent **fault scarp** is often associated with active fault lines and may be visible on topographic maps and HGM layers. Look for linear scarps or abrupt elevation changes in areas of suspected lineaments.

  - **Tectonic Landforms**: **Ridges, valleys**, or **faulted terrain** that align with linear features (e.g., faults or fractures) are direct indicators of lineaments.

  - **Landform Patterns**: In areas of tectonic activity, **parallel ridges** or **valleys** that run along straight lines can indicate the presence of lineaments controlling landscape development.

- **Field Application**: Field teams can use HGM maps to identify locations where faults or fractures might be influencing geomorphological features. These areas can then be mapped on the ground, with detailed field surveys confirming the lineament presence.

C. Analysing Groundwater Flow and Recharge Zones

- **Method**: Faults and fractures can act as conduits for groundwater flow, altering subsurface hydrology. Groundwater flow can be traced using **hydrogeological data** (e.g., aquifers, recharge areas) often incorporated in HGM maps.

- **How to Detect Lineaments:**
  - **Perpendicular to groundwater flow**: Lineaments often intersect groundwater flow lines, influencing the direction of water movement. In HGM maps, look for areas where groundwater flow is disrupted or directed along linear features.
  - **Springs and Wells**: Springs or **artesian wells** may occur along lineaments due to faulting or fracturing that allows groundwater to surface. These can be mapped using **hydrological data** on HGM maps.
  - **Groundwater Recharge Zones**: Faults or fractures can act as zones of **increased permeability**, leading to **higher groundwater recharge**. Areas of linear high permeability could correlate with underlying lineaments.
- **Field Application**: Field teams may use HGM maps to locate areas with groundwater anomalies, like springs or recharge zones, which can be examined to determine whether the water pathways correspond to structural lineaments.

D. Soil and Sediment Characteristics

- **Method**: Lineaments can affect the distribution of **soil types** and **sediment characteristics**. For example, fault zones may have distinct soil characteristics due to tectonic activity (e.g., soil erosion, fault gouge).
- **How to Detect Lineaments:**
  - **Soil Disturbance**: Faulting or fracturing often leads to the formation of distinct soil types along lineaments, such as **fractured rock debris** or **fault gouges.**

- ○ **Sediment Distribution**: In river valleys or sedimentary basins, **sediment deposition patterns** may align with lineaments, especially in fault-controlled environments (e.g., sediment accumulation in fault-bounded basins).

- **Field Application**: Field surveys can be conducted in areas where **soil anomalies** or **sediment patterns** appear aligned with linear features on HGM maps. Soil samples can help confirm the presence of structural features like faults or fractures.

E. Faults and Tectonic Structures

- **Method**: Direct identification of **faults** on HGM maps, often represented as linear features or zones of displacement, can reveal lineaments associated with tectonic activity.

- **How to Detect Lineaments**:

  - ○ **Fault Zones**: Look for **linear fault zones** or fractures that may correlate with lineaments on the map. These can often be traced visually along stream valleys, scarps, or ridges.

  - ○ **Displacement Patterns**: In areas with active tectonism, **offset landforms** such as displaced river channels, fault scarps, or tilted strata are associated with lineaments.

- **Field Application**: In the field, lineament features identified through fault zones or displacement patterns on HGM maps can be further explored to identify active faulting or evidence of tectonic movement.

5.2.3. Steps to Analyse Lineaments Using HGM Maps in the Field

I. Gather Relevant HGM Data Layers

- Collect **topographic**, **hydrological**, **geomorphological**, and **soil maps** for the region of interest.

- **Overlay drainage patterns, landforms, groundwater flow**, and **fault zones** to identify potential areas with linear features.

II. Preliminary Lineament Identification

- Use HGM maps to identify straight, linear features in **drainage patterns, elevation changes**, and **soil anomalies** that may indicate lineaments.

- Focus on areas with potential **fault zones, tectonic landforms**, and **disrupted drainage** patterns.

III. Field Verification

- Navigate to areas identified on the HGM maps as having linear features (e.g., fault scarps, linear valleys, or groundwater anomalies).

- **Ground truthing**: Inspect physical signs of lineaments on the ground such as visible fault scarps, fractures in rock, or distinct geomorphic features that correlate with map data.

IV. Geological Survey

- Conduct geological surveys in the identified lineament zones to observe structural features such as fault planes, offset landforms, or fractures.

- **Sample soils or sediments** in suspected lineament zones to confirm whether geological activity is present.

V. Refine Lineament Mapping

- Once verified, use **GPS coordinates** to mark the exact location of lineaments in the field and update maps accordingly.

- Continue using GIS tools to refine lineament mapping based on field observations.

5.3. The Map Pictures

5.3.1. Practical Example: Using HGM Maps to Identify Lineaments

Let's say you are tasked with analysing **lineaments in a fault zone** in a mountainous area:

1. **HGM Map Analysis:**

    - You observe that the **drainage pattern** in the area is **rectilinear**, with streams flowing along straight paths, which may indicate fault-controlled topography.

    - **Fault scarps** are visible in a topographic map overlay, and **springs** are clustered along a linear feature in the **hydrological map**, suggesting groundwater movement along a fault.

2. **Field Verification:**

    - In the field, you visit the areas where the fault scarps are located. You find **displaced layers of rock**, visible **fault planes**, and **fractured rock** that confirm the presence of the lineament.

3. **Conclusion**: You confirm that the linear features observed in the HGM map are indeed related to a fault line and map this lineament with greater accuracy for further geological analysis.

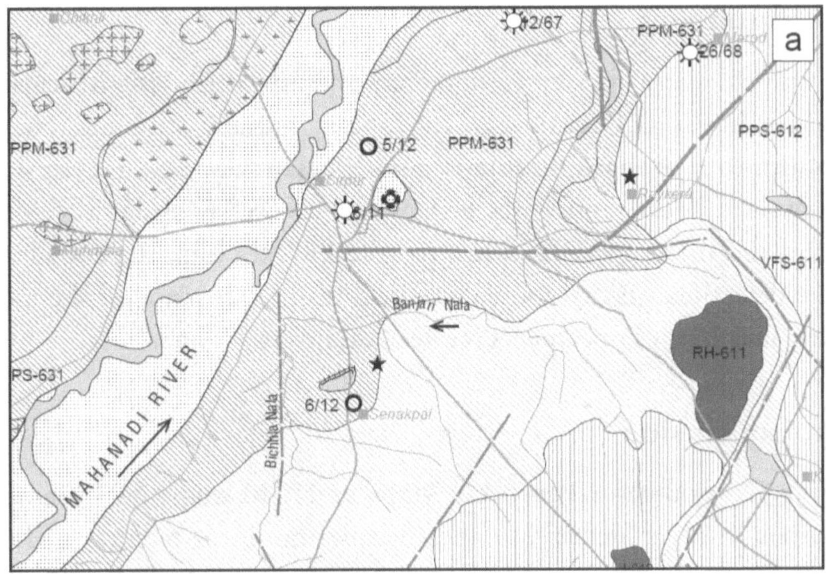

Water Prospects Map

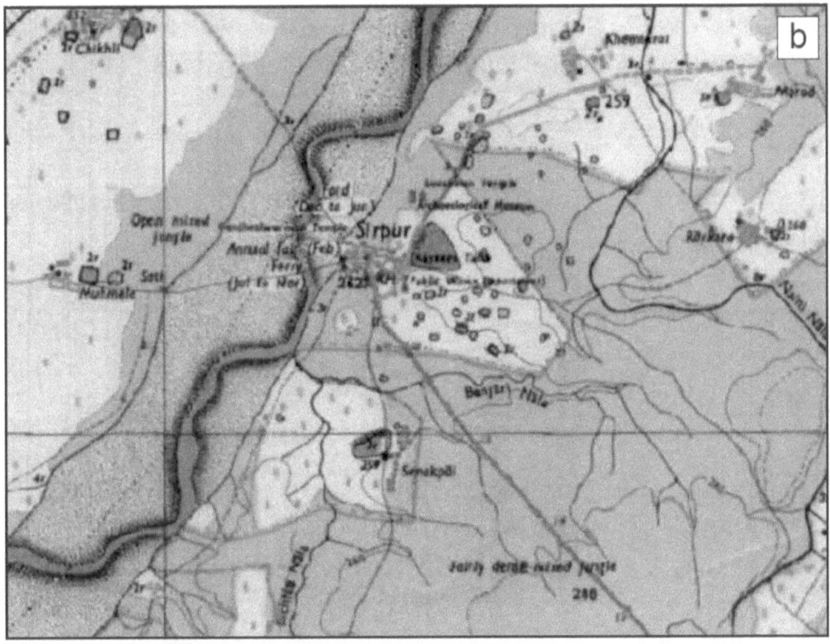

*Fig. 4 Sirpur, Chhattisgarh (showing general arrangement only)
(In Courtesy of Manual of RGNDW User Manual)*

5.4. Summary

In the field, these map representations help in the **initial identification** and **verification** of lineaments. The process typically combines **remote sensing data**, **topographic maps**, **geological surveys**, and **geophysical analyses** to locate and verify lineaments. By overlaying multiple data layers and using specialised mapping techniques (e.g., slope analysis, edge detection, gravity anomalies), field teams can pinpoint the most likely locations of lineaments for detailed investigation.

5.5. Conclusion

Using **Groundwater Prospects Maps or HGM maps** for lineament analysis involves identifying and interpreting linear features such as drainage patterns, geomorphological structures, and hydrological data that indicate underlying structural features like faults and fractures. By overlaying these data layers and conducting field verification, you can identify, map, and analyse lineaments that may have important implications for geological studies, groundwater analysis,

◆　◆　◆

The Diviner's Toolkit: 10 Methods for Tapping Lineament-Linked Groundwater

10 Divining Practices to identify the Location of Borewell Drilling

6.1. Divining with Physically Visualising Lineaments in the Field

6.2. **Y-rod Divining: Understanding the Technique**

6.3. L-Rods (Oriented Rods): An In-Depth Exploration

6.4. Pendulums: A Timeless Tool of Divination and Discovery

6.5. Zinc or Copper Rod Divining: Harnessing the Energy of Metal Rods

6.6. Dowsing Boards: A Guided Path to Divination

6.7. **Water witching:** The Two-Stick Method

6.8. Dancing Coconut Divining: A Detailed Exploration

6.9. **Methods of Change in Temperature**

6.10. **Others divining:** The Pot or Mug Method, Eggs Method, etc.

◆ ◆ ◆

6.1. Divining with Physically Visualising Lineaments in the Field

Beneath our feet and across the landscapes we traverse lies a hidden language etched into the Earth's surface—lineaments. These subtle, often elusive features, carved by tectonic forces, erosion, and time itself, hold the key to understanding the geological and hydrological secrets of the land. While maps and satellite imagery offer glimpses of these patterns, there is nothing quite like standing in the field, tracing these features with your own eyes and hands, to truly grasp their significance.

This chapter embarks on a journey of discovery, guiding readers through the art and science of physically visualising lineaments in their natural settings. From identifying fault traces and fracture zones to interpreting their geological and environmental implications, we delve into the techniques and tools that transform abstract maps into tangible realities.

By blending field observation with scientific interpretation, we unlock the stories embedded in Earth's linear features—stories that reveal the forces that have shaped our planet and continue to influence its evolution. Whether you're a seasoned geologist or a curious explorer, this chapter will sharpen your vision and deepen your understanding of the landscape's hidden geometry.

Detecting lineaments visually in the field by observing the **topographic trends** of the land is a skill developed through experience and understanding how geological features manifest on the surface. Lineaments, which are linear or slightly curved geological features such as faults, fractures, joints, and folds, often show clear expressions in the **landforms** and **terrain** that can be detected with the naked eye.

Here's a step-by-step guide to detecting lineaments through visual observation in the field, focusing on physical and topographic trends:

6.1.1. Look for Straight or Linear Features

- **Linear Landforms**: The most apparent surface expression of a lineament is a **straight or slightly curved feature** in the landscape. This could be in the form of:

    - **Ridges**: Straight or gently curving ridges or mountain chains may indicate fault lines or structural alignments.

    - **Valleys**: Linear valleys or depressions, especially those running parallel to ridges or faults, are often formed by faulting or other tectonic activities.

    - **Stream Channels**: River or stream channels that follow straight or rectilinear paths can be indicative of underlying geological structures like faults or fractures that control the flow of water.

How to Identify

- Walk through the area and keep an eye out for **straight ridgelines** or **linear valleys** that might stand out against the natural curve of the terrain.

- Look for any unusual **displacement** in the landscape where **topography abruptly changes** along a straight line.

Example: In a mountainous region, a **ridge** running straight for several kilometres could be the result of faulting. If the same ridge is aligned with a series of small **streams** or **valleys**, it suggests that the lineament is tectonically controlled.

6.1.2. Check for Sudden Elevation Changes

- **Topographic Changes**: Sudden changes in **elevation**, such as steep **scarps, cliffs**, or **fault scarps**, often indicate the presence of lineaments due to faulting, folding, or other structural deformation.

- A **scarp** is a **steep cliff** or **slope** that forms due to fault displacement, and it is often a visible sign of a lineament.

- **Raised Landforms**: Look for **uplifted blocks** or **subsided areas**, which often correlate with tectonic movement along fault lines.

- A straight scarp in an area where the surrounding land is generally flat or gently sloping could be indicative of a **recent fault** or **fracture zone**.

How to Identify

- Look for **linear cliffs or scarps** where the landscape abruptly drops or rises. These features could be on the surface or in areas where you notice a large displacement in the landforms.

- A **lineament** may present itself as a **geological feature** that disrupts the **natural curvature** of the terrain, such as a straight or nearly straight edge on a ridge or hill.

Example: In an area with rolling hills, you notice a **linear cliff** that runs across a flat valley floor. This sudden change in elevation could point to a fault or lineament zone where tectonic forces have caused a vertical displacement.

6.1.3. Observe River and Drainage Patterns

- **Straight Drainage**: The flow of rivers or streams can reveal lineaments, especially when they follow straight or **angular paths. Linear river courses** are often controlled by underlying **faults** or **fractures** in the bedrock.

  - A **displaced river** or **valley** that follows a **straight line** or **sharp angle** may indicate that the river is following a fault or fracture zone.

- ○ **Rectangular or angular patterns** in the drainage system are often associated with **tectonic faults** that cause streams to change direction abruptly.

How to Identify

- Stand on higher ground and scan the landscape for rivers or streams that appear to flow in **straight lines** or follow **angular bends**.

- Look for **river courses** that are aligned with **ridge lines** or **valleys**, as these are often controlled by faults.

Example: A river flowing **directly along a straight line** for several kilometres and then suddenly turning at a sharp angle could indicate that the water is following a **fault zone** or **fracture**.

6.1.4. Look for Erosion or Soil Differences

- **Differential Erosion**: Lineaments can create **differential erosion** due to the contrasting resistance of rock types along fault lines. For example:

  - ○ **Harder rocks** (like granite) often form **ridges** or **hills**, while **softer rocks** (like shale or sandstone) may form **valleys** or **depressions**.

  - ○ Faults and fractures can create zones where the landscape erodes in a straight or linear fashion, often resulting in **sharp ridges**, **valleys**, or **terraces** that follow a linear path.

How to Identify

- Scan for areas where the **erosion patterns** create distinct **linear ridges** or valleys that run parallel or perpendicular to the surrounding terrain.

- **Uneven soil types**: In faulted areas, soil types may change abruptly, which can be visible on the ground as **changes in vegetation** or **surface soil**.

Example: In a region with alternating layers of hard and soft rock, you may observe that the **softer rock** erodes faster and creates a **linear valley**, while the **harder rock** forms a **ridge** along the fault line.

6.1.5. Examine Vegetation and Soil Patterns

- **Vegetation / Tree's Alignment**: Certain vegetation types and patterns can be seen more prevalent along a series of bending of trees towards fault lines or fractures due to differences in soil and water conditions caused by underlying lineaments.

  - **Fault Zones**: Areas along faults or fractures may have different soil properties (e.g., higher permeability or moisture retention), which can affect **vegetation growth**.

  - **Fracture-Related Features**: Lineaments may show up as **linear patterns of vegetation** that are aligned with faults, fractures, or other geological features.

How to Identify

- Look for **linear bands** of vegetation that follow a straight or curved path, potentially indicating **structural control** of underlying features.

- Areas with **different vegetation densities** along straight features or ridgelines may suggest a change in soil type due to faulting or fracturing.

Example: A line of trees or **dense vegetation** along a straight path bending towards a forest could point to a fault or fracture zone where the underlying conditions are favourable for plant growth due to the presence of groundwater or nutrients.

6.1.6. Check for Geomorphic Features like Fault Scarps or Ruptures

- **Fault Scarps: Fault scarps** are often the most obvious surface feature indicating a lineament. These are **steep cliffs** or **slope breaks** caused by **vertical displacement** along a fault line.

    - **Linear Faults:** Fault scarps may form along a **straight line**, creating distinct boundaries between different rock types or landforms.

- **Rupture Zones:** Large-scale **land ruptures** can occur along faults, creating visible **disruptions** in the landscape such as cracked ground, offset streams, or displaced roads.

How to Identify

- Walk along ridges or valleys and look for any **steep linear cliffs** or **sudden changes in elevation** that follow a straight line.

- **Examine cracks or fissures** in the ground along linear features, which may indicate a fault or fracture.

Example: While hiking through a hilly area, you come across a **sudden vertical cliff** with **displaced rock strata** that run along a straight line. This is likely a fault scarp caused by tectonic movement along the fault line.

6.1.7. Look for Horizontal or Vertical Displacement in Rock Layers

- **Strata Offset:** If you come across exposed rock outcrops, look for **offset rock layers** or **displaced strata** along a straight or curved line. This can indicate that the ground has been **shifted horizontally** or **vertically** along a fault or fracture zone.

How to Identify

- **Examine rock formations** and **strata** to see if they align smoothly or are **offset** by a fault line.

- **Vertical displacement** is often easy to spot as a clear **step-like feature** in the rock layers.

Example: While observing an exposed rock face in a riverbed, you notice that two layers of rock that should be aligned are **displaced vertically**, forming a **step** along the fault line. This is a direct indication of a fault, which could be part of a larger lineament.

Fig. 5 (up) and 6 (down), Village Gabhoda, water well location, situated on lineament stream line. Visualisation traced by the expert team. Another picture below shows the search for a tubewell site.

6.1.8. Case Study 3: Open Visualisation of Lineaments for Shallow Open Wells in Gabhoda Village, Block Bodla, District KabirDham (Chhattisgarh, India)

I. Introduction

Access to safe and sustainable drinking water is a critical need in rural areas like Gabhoda village, located in the Bodla block of KabirDham district. The region experiences challenges in securing potable water, especially during dry seasons. The use of geological and hydrogeological techniques, such as lineament mapping, provides an innovative solution for identifying potential water sources. This case study outlines how lineaments passing through valley streams were utilised to locate and establish shallow open wells for drinking water purposes in Gabhoda village.

Geographical and Hydrogeological Context

- **Village**: Gabhoda

- **Block**: Bodla

- **District**: KabirDham

- **Topography**: The region features undulating terrain with valleys, streams, and forest cover

- **Hydrogeology**: Lineaments (faults, fractures, and joints in the rock) are key conduits for groundwater recharge and flow. These lineaments often align with valleys and streams where groundwater accumulation is higher.

II. Objectives

1. To identify suitable locations for shallow open wells using lineament visualisation techniques.

2. To ensure sustainable water availability for the villagers by targeting zones of high groundwater potential.

3. To reduce dependency on distant or unreliable water sources.

III. Methodology

1. Lineament Mapping

- Remote sensing and GIS techniques were used to identify lineaments. Satellite images and topographic maps were analysed to delineate fractures, faults, and joints intersecting the village area.

- Lineaments coinciding with streams in valley regions were given priority, as these are natural recharge zones.

2. Ground Validation

- Field surveys were conducted to validate the mapped lineaments.

- Geophysical techniques, such as electrical resistivity surveys, were used to confirm groundwater potential along the identified lineaments.

3. Well Digging

- Shallow open wells were dug at strategic points along the validated lineaments.

- Construction ensured minimal ecological disruption while maintaining structural integrity.

4. Community Engagement

- Villagers were involved in the planning and construction processes to ensure ownership and sustainability.

- Training sessions on well maintenance and water conservation were conducted.

IV. Results

1. **Well Yield**: The shallow open wells provided consistent water yield even during dry months, as they tapped into lineament-fed aquifers.

2. **Water Quality**: Preliminary tests indicated that the water was safe for drinking after basic filtration.

3. **Community Impact**: Over 80 households benefited from the initiative, reducing the burden of fetching water from distant sources.

4. **Environmental Sustainability**: The project had minimal environmental impact and enhanced awareness of groundwater conservation among villagers.

V. Challenges

- Initial resistance from the community due to lack of awareness.

- Technical constraints in mapping and interpreting complex lineament systems.

- Seasonal variations in water table levels required ongoing monitoring.

Conclusion

The integration of lineament mapping with community participation proved to be an effective approach to addressing drinking water scarcity in Gabhoda village. By leveraging natural hydrogeological features, sustainable and reliable water sources were created for the community. This case study highlights the potential for similar interventions in other water-scarce regions.

Recommendations

1. **Periodic Monitoring**: Regular monitoring of well yields and water quality to ensure long-term sustainability.

2. **Capacity Building**: Continued community education on water management practices.

3. **Replication**: Expansion of the approach to other villages in KabirDham district facing similar challenges.

This project demonstrates how modern techniques combined with local knowledge can make a tangible difference in rural water resource management.

Fig. 7 Y-rod Divining Demonstration Picture

6.2. Y-rod Divining: Understanding the Technique

I. Introduction to Y-rod Divining

Y-Rod divining, also known as "water witching," is one of the most widely recognised forms of divining or dowsing. It involves the use of a Y-shaped branch or rod, typically made from a flexible wood like willow, hazel, birch or even metal, to locate water sources underground. The diviner holds the two arms of the Y-rod in their hands, with the stem pointing forward, and walks over a suspected groundwater source. When the diviner passes over a water vein or underground stream, the rod is believed to involuntarily bend or "react," signalling the presence of water below the ground.

Symbolism

- The forked Y-rod is symbolic in many cultures. In Europe, it was thought that **hazel** wood (often used for Y-rods) had magical properties, and it was associated with fertility and life-giving forces, making it an ideal material for water divining.

- In other cultures, the Y-shaped branch is believed to act as a kind of antenna, picking up subtle vibrations or energies from the Earth.

II. How Y-rod Divining Works: The Mechanics

The basic principle of Y-rod divining rests on the idea that the rods respond to subtle changes in the environment, particularly the electromagnetic fields generated by underground water. Here's how the process works:

1. **Choosing the Rod**: The Y-shaped rod can be made from various materials, though traditional practitioners favour wood for its sensitivity. The rods are often cut from specific tree species, such as hazel, ash, or willow, though others like brass or copper have also been used in modern practices.

2. **Positioning the Rod**: The diviner grips the two arms of the Y with their hands, while the stem points forward. The diviner typically holds the rod at a comfortable angle, with their arms slightly apart to allow the rod to move freely.

3. **The Walk**: The diviner begins to walk slowly over the area where they believe water may be located. As they move across the land, they focus their attention on the rod and remain attuned to any subtle movements.

4. **The Reaction**: When the diviner passes over an area where groundwater is believed to be present, the rod will respond. The common reactions include:

   ○ **Bending or Tilting**: The Y-rod may bend sharply downwards or to one side, indicating the location of the water.

   ○ **Swinging or Rotating**: In some cases, the Y-rod may spin, move back and forth, or twist, suggesting a potential source of underground water.

5. **Reading the Signs**: Once the rod has moved, the diviner interprets the direction and intensity of the movement to determine where to drill, the depth of the water source, or the flow direction of an underground stream.

III. Scientific Explanations and Debates

While Y-rod divining has been practised for centuries, its efficacy remains a subject of debate. Scientists and sceptics often attribute the movements of the rod to the **ideomotor effect**, which is the subconscious movement of the hands and arms. Essentially, the diviner may unknowingly cause the rod to move based on their thoughts and beliefs, rather than external influences.

Electromagnetic Fields (EMF) Hypothesis: Some proponents of Y-rod divining suggest that underground water sources emit electromagnetic fields that can affect the rods. Water, especially when flowing through porous or mineral-rich rock formations, can generate electrical charges that might be detectable by sensitive dowsers. While there is no definitive scientific proof to support this claim, a few studies have attempted to explore the possibility.

Scientific Studies and Scepticism: In a series of controlled studies, diviners were asked to locate water sources without prior knowledge of their locations. In some cases, diviners did accurately pinpoint the location of groundwater, but in other instances, the results were inconsistent. A **1978 study** by the National Research Council found that diviners performed no better than chance when tested under controlled conditions. However, diviners have pointed out that successful results often occur in natural, unregulated settings where they are familiar with the landscape and can apply their intuition, which is hard to replicate in lab experiments.

IV. Real-World Examples of Y-rod Divining Success

Despite scepticism, there are numerous anecdotal reports and case studies where Y-rod divining has been credited with success, especially in rural or agricultural settings.

1. **Case Study: Agricultural Water Sourcing in the U.S.** In the early 20th century, many farmers in the American Midwest relied on dowsing to locate water sources for irrigation. One notable case occurred in the 1930s when a group of farmers in **Kansas** used Y-rod divining to locate underground water during a severe drought. According to reports, the diviners were able to pinpoint areas where they subsequently drilled successful wells that provided a reliable water supply for crop irrigation.

2. **Case Study: Water Supply for a Village in India** In rural India, dowsing has been used by local communities to locate groundwater sources for drinking water. In one case, a village

in **Rajasthan** was suffering from acute water shortages. A local dowser was called in, and using the Y-rod, they successfully located a subterranean stream beneath a dry part of the village. A well was drilled at the location, and it provided a freshwater source for the village for years to come.

3. **Case Study: Locating Underground Streams for Mining Operations** In certain mining operations, diviners have been used to locate water sources that could affect mining processes. In the 1980s, a mining company in **Nevada, USA**, reportedly used dowsers to identify areas where underground water could disrupt excavation. Their results, according to reports, helped avoid costly water-related issues and improved the efficiency of the mining operation.

V. Statistics and Usage in Modern Context

While scientific validation of Y-rod divining remains elusive, its use continues, particularly in rural areas or where access to expensive geophysical surveying equipment is limited. According to a **2015 survey of U.S. dowsers** conducted by the American Society of Dowsers, over 60% of diviners surveyed had successfully located water for well drilling, and approximately **80% of dowsers** reported using their skills regularly in agricultural or land management contexts.

In a **2018 survey in India**, over **40,000 rural farmers** were interviewed, and it was found that more than **25%** relied on traditional methods, including Y-rod divining, to locate water for wells. This trend is particularly common in rural regions with limited access to technology and where divining has deep cultural significance.

VI. Limitations and Challenges of Y-rod Divining

Despite its long history and widespread use, there are limitations to the practice:

- **Inconsistency of Results**: Divining is often highly subjective, with results varying based on the experience of the dowser, the environment, and even the specific tools used.

- **Environmental Interference**: Certain soil types, minerals, and surface conditions may interfere with a dowser's ability to detect underground water sources.

- **Lack of Scientific Consensus**: There is no universally accepted scientific explanation for why or how Y-rod divining works, leaving it on the fringes of mainstream water exploration techniques.

VII. Practical Tips for Using Y-rod Divining

1. **Choose the Right Rod**: For beginners, a flexible wooden Y-rod is recommended. The wood should be cut fresh, with the ends of the "arms" roughly 12 to 18 inches long.

2. **Stay Focused and Calm**: It's important for the diviner to remain relaxed and focused during the session. Any distractions can affect the outcome.

3. **Interpret with Care**: The movement of the rod should be interpreted with caution. A single, slight bend doesn't always indicate water; it could be the result of environmental conditions or subconscious movements.

Fig. 8 L-Rod Divining Demonstration pics

6.3. L-Rods (Oriented Rods): An In-Depth Exploration

I. Introduction to L-Rods

L-Rods, also known as oriented rods, are a traditional dowsing tool used to detect energy fields, locate underground water sources, and identify ley lines or buried objects. These rods are typically fashioned in the shape of an "L," made from materials like copper, brass, or steel, and are held loosely in each hand. Their simplicity, versatility, and historical significance have cemented their place in both ancient and modern divination practices.

II. Structure and Design

The L-Rod consists of two components:

1. **Handle**: The shorter part of the "L," typically 4–6 inches long, held in the hand.

2. **Rod**: The longer part, usually 12–20 inches, extends outward and moves freely.

The rods are often balanced to allow sensitivity to subtle movements. Some L-Rods include sleeves or bearings on the handles to minimise friction, enhancing their responsiveness.

III. Method of Use

1. **Preparation**: The dowser clears their mind and sets an intention, such as finding water or sensing energy lines.

2. **Positioning**: Standing with feet shoulder-width apart, the user holds an L-Rod in each hand, parallel to the ground and pointing forward. The rods should move freely without intentional force.

3. **Movement**: As the dowser walks slowly, the rods may cross, swing outward, or exhibit other movements at specific points. These movements are interpreted as responses to energy or the presence of the sought object.

4. **Interpretation:**

   - **Crossing rods** often indicate the target (e.g., water or an object).

   - **Outward movements** may signal boundaries or avoidance.

Practitioners recommend working in quiet, distraction-free environments to enhance sensitivity.

IV. Advantages of L-Rods

- L-rods are seen as more **sensitive** than Y-rods and are often preferred for more precise measurements or locating smaller sources of water.

- The movement of the rods is believed to be involuntary, as they cross when the dowser unknowingly responds to energy changes in the environment.

V. Symbolism and Cultural Significance

L-Rods hold both practical and mystical significance. They are often viewed as tools that bridge the physical and metaphysical realms, symbolising humanity's ability to interact with unseen forces. In folklore, dowsers were seen as intermediaries with divine or natural energies.

The shape of the L-Rod itself is symbolic:

- The **"L"** represents balance and alignment between the physical direction (the handle) and the spiritual or energy field (the rod).

- In some cultures, L-Rods are imbued with protection symbolism, believed to shield dowsers from negative influences.

VI. Examples of Application

1. **Water Divination**: Farmers and landowners have historically used L-Rods to locate underground aquifers for wells. Studies in rural areas suggest dowsers achieve moderate success rates, sometimes as high as 70%, in areas with known water deposits.

2. **Energy Mapping**: Some practitioners use L-Rods to locate energy vortices or ley lines, aligning these with ancient structures like Stonehenge or the Great Pyramids.

3. **Paranormal Investigations**: In modern times, L-Rods are occasionally used to detect shifts in energy fields during ghost hunts or spiritual explorations.

V. Statistics and Effectiveness

While scepticism exists, anecdotal evidence supports the practical use of L-Rods:

- A 1995 German study found a 75% accuracy rate among experienced dowsers in controlled environments with hidden water sources.

- In agricultural applications, rural surveys show that 60–80% of respondents consider dowsing (including L-Rods) a reliable method for locating water.

Modern scientific consensus attributes the movements of L-Rods to the ideomotor effect—subtle, unconscious muscle movements influenced by the dowser's expectations. However, many practitioners argue for an unexplained connection to natural energy fields.

VI. Crafting and Customisation

Enthusiasts often create their own L-Rods, adding personal touches like inscribed symbols, crystal embellishments, or ergonomic handles. Copper is the preferred material due to its conductive properties, believed to enhance sensitivity to energy fields.

6.3.1. Case Study 4: Borewell Site Selection Using L-Rods at Village Ghorbhatti, Block Abhanpur

Background

Water resource management is a critical aspect of rural infrastructure development under JJM (JAL JIVAN MISSION). In Village Chorbhatti, Block Abhanpur, a borehole was required near the overhead water tank to meet the community's water needs effectively. Selecting an appropriate site for the borehole was essential to ensure adequate water supply and efficient system integration.

Objective

The primary objective of this project was to identify an optimal location for the borehole that ensures reliable water availability while minimising costs associated with infrastructure and resource utilisation.

Approach and Methodology

1. Site Selection Criteria

The site needed to be:

- Close to the overhead water tank to minimise water pumping distance and associated energy costs.

- Located where sufficient underground water reserves could be tapped sustainably.

2. Divining Technique Using L-Rods

The project employed the L-Rod divining method, a traditional technique for detecting underground water sources. This method involves using L-shaped rods to sense changes in electromagnetic fields, which are believed to indicate the presence of water.

3. Expert Involvement

Shri P.S. Rana, a hydrologist with expertise in water resource management, was engaged to conduct the divining process. His experience and skill ensured the reliability of the results and helped in validating the selected site.

Implementation and Findings

- **Process Execution:** The divining process was carried out under the guidance of Shri P.S. Rana. Multiple locations around the overhead tank were examined to ensure comprehensive coverage of the area.

- **Site Finalisation:** Based on the divining results, a site was selected where underground water availability was deemed optimal. The location met the criteria of proximity to the overhead tank and potential water yield.

Outcomes

1. **Technical Viability:** The chosen site demonstrated favourable conditions for borewell drilling.

2. **Cost-Effectiveness:** By selecting a site near the overhead tank, the project minimised infrastructure costs related to water pumping and distribution.

3. **Sustainable Resource Use:** The divining process helped ensure the borewell would tap into a sustainable water source, avoiding overexploitation.

◆ ◆ ◆

Fig. 9 Divining work by Shri P. S. Rana, Hydrologist,
through L-Rod Method.

6.3.2. Case Study 5: Borehole Site Selection at Tourism Motel, Village Tumribod, District Rajnandgaon, Chhattisgarh

I.Background

Tourism development projects in rural areas often require reliable water sources to support infrastructure and visitor needs. At a tourism motel site in Village Tumribod, District Rajnandgaon, a borehole was necessary to provide a sustainable water supply for the facility's operations. Proper site selection was critical to ensure water availability and minimise operational challenges.

II. Objective

The objective was to identify a suitable location for a borehole that could meet the water requirements of the tourism motel efficiently and sustainably.

III. Approach and Methodology

1. Site Selection Criteria

The site selection process focused on:

- Proximity to the tourism motel to ensure ease of water distribution.

- Identifying a location with sufficient underground water reserves for long-term usage. While already 4-5 bore wells failed earlier.

2. Divining Technique Using L-Rods

As in the previous project, the L-Rod divining method was employed to detect underground water sources. This traditional technique relies on changes in electromagnetic fields sensed by L-shaped rods to locate water.

3. Expert Involvement

Hydrologist Shri P.S. Rana, known for his expertise in water resource management and divining, was engaged to conduct the site selection process. His professional experience ensured the reliability of the results.

IV. Implementation and Findings

- **Divining Process:** Shri P.S. Rana conducted the divining process across the motel site. Multiple locations were examined to ensure a comprehensive analysis and selection of the most viable water source.

- **Site Finalisation:** A site with promising underground water potential was identified near the tourism motel. The selected location met all necessary criteria, including ease of access and projected water yield.

- **Successful Borewell Drilling:** The borewell was successfully drilled at the identified site, confirming the accuracy of the divining process.

V. Outcomes

- **Reliable Water Source:** The borewell provided a sustainable water supply, fulfilling the motel's operational needs.

- **CostEfficiency:** The divining method minimised the cost of exploratory drilling by accurately identifying the optimal site.

- **Enhanced Tourism Support:** The availability of a dependable water source bolstered the functionality of the tourism motel, enhancing its appeal to visitors.

◆ ◆ ◆

6.4. Pendulums: A Timeless Tool of Divination and Discovery

I. Introduction to Pendulums

A pendulum is a simple yet powerful tool used for divination, spiritual guidance, and decision-making. It consists of a small, heavy object, often made from materials such as crystal, metal, or wood, suspended from a chain or string. Despite its simplicity, the pendulum is revered for its ability to tap into subconscious insights and energies, bridging the physical and spiritual realms.

II. Structure and Types of Pendulums

Pendulums come in various shapes and materials, each chosen for specific purposes:

- **Crystal Pendulums**: Made from gemstones like amethyst, quartz, or obsidian, these are prized for their unique energetic properties.

- **Metal Pendulums**: Often made from brass or copper, these are favoured for their precision and neutrality.

- **Wooden Pendulums**: Ideal for grounding work and those seeking a natural connection.

- **Chamber Pendulums**: Feature a hollow interior to hold a small item, such as essential oil, herbs, or water, aligning with the intent of the session.

The pendulum's design often includes a pointed or rounded tip, enhancing its responsiveness and accuracy during use.

III. How It Works

- The dowser holds the chain or string of the pendulum between their fingers, allowing the pendulum to hang freely.

- As the pendulum swings, the dowser is attuned to the subtle movements, interpreting changes in direction as indicators of underground water, minerals, or even specific properties of the water source.

- A **"yes" or "no"** question may be asked to determine whether the pendulum swings in a particular direction (such as clockwise or counterclockwise). The dowser interprets these swings to find water or other desired substances.

IV. Method of Use

Pendulum dowsing involves a straightforward process that requires focus and intention:

1. **Preparation**

   - Choose a pendulum that resonates with you.

   - Cleanse it physically and energetically to ensure neutrality.

   - Establish a connection by holding it and setting your intention.

2. **Calibrating Responses**:

   - Hold the pendulum steady, suspending it above a flat surface.

   - Ask simple yes-or-no questions to determine how the pendulum moves for «yes,» «no,» and «maybe.» Typical movements include:

     - **Circular motion**: Often signifies "yes."

     - **Linear motion** (back and forth): Often signifies "no."

     - **Side-to-side motion** or lack of movement: Indicates "uncertain" or "maybe."

3. **Dowsing Session**:

   - Pose specific questions or focus on a topic.

   - Interpret the pendulum's movements as it responds to subtle energetic influences or subconscious guidance.

   - For advanced use, combine with charts or maps for locating objects, answering complex queries, or assessing chakras.

V. Symbolism and Spiritual Significance

The pendulum symbolises the connection between the physical and the metaphysical:

- **Balance and Harmony**: Its swinging motion represents equilibrium and the interplay of opposing forces.

- **Guidance and Intuition**: As an extension of the user's subconscious, it embodies trust in inner wisdom.

- **Spiritual Connection**: Many traditions view the pendulum as a conduit for divine energies or higher consciousness.

Its use transcends cultures and eras, appearing in practices like ancient Egyptian divination, medieval dowsing, and modern Reiki healing.

VI. Examples of Pendulum Applications

1. **Decision-Making**: Individuals consult pendulums for guidance on personal, financial, or relationship matters.

2. **Energy Work**: Practitioners use pendulums to assess and balance chakras, ensuring energetic alignment.

3. **Locating Objects**: Dowsers use pendulums over maps to identify the location of lost items, water sources, or minerals.

4. **Health and Wellness**: Alternative healers employ pendulums to diagnose imbalances or suggest remedies.

VII. Statistics and Modern Relevance

While the pendulum's efficacy is challenging to measure scientifically, its widespread use suggests enduring appeal:

- Surveys show that **30-40% of alternative healing practitioners** incorporate pendulums in energy work or spiritual counselling.

- Anecdotal evidence suggests accuracy rates of **60-80%** for experienced users when asking calibrated questions.

- Studies on ideomotor responses—the unconscious muscle movements that influence pendulum swings—support its psychological grounding.

Despite scepticism, the pendulum remains a respected tool in both spiritual and psychological exploration.

VIII. Choosing and Caring for a Pendulum

To optimise results, users should:

1. **Select a Pendulum**: Choose based on material, shape, or intuitive appeal.

2. **Cleanse and Recharge**: Use methods such as moonlight, smudging, or placing on a selenite slab.

3. **Store Properly**: Keep in a soft pouch to protect the material and maintain its energy.

Final Thoughts

The pendulum's simplicity belies its profound impact. By serving as a tangible link between the conscious mind and deeper intuition, it continues to empower users across cultures and disciplines. Whether viewed as a mystical instrument or a psychological tool, the pendulum's resonance in human history and spirituality is undeniable. Pendulums are often considered especially useful for **fine-tuned questions** or when the dowser wants to obtain more detailed information, such as the depth or quality of the water.

6.4. Zinc or Copper Rod Divining: Harnessing the Energy of Metal Rods

I. Introduction to Straight Rod Divining

Straight rods, often made of zinc or copper, are versatile tools used in the ancient art of dowsing. These rods are prized for their conductive properties, believed to enhance their sensitivity to subtle energy fields. Unlike L-Rods or Y-Rods, straight rods are a minimalist tool, symbolising the direct connection between the dowser and the natural energies they seek to interpret.

II. Structure and Material

Straight rods are simple, slender lengths of metal, typically 10 to 20 inches long, and are often polished for a comfortable grip. They may include slight bends at one end to aid in handling or may be entirely straight for unimpeded motion.

- **Copper**: Known for its high conductivity, copper rods are widely used for detecting energy fields and water. Copper's metaphysical properties are associated with enhancing intuition and amplifying energy.

- **Zinc**: Often combined with copper, zinc rods are believed to balance energies. They are particularly favoured for locating groundwater due to their supposed affinity with Earth energy.

The material choice often depends on the dowser's personal preference, the environment, and the intended purpose.

III. How It Works

- As the dowser walks, the rods are held loosely with each hand. The rods are said to cross over one another when they pass over a source of water or minerals, or they might swing apart.

- The movement of the rods is interpreted similarly to the L-Rod, with the crossed position suggesting a location of interest.

- Some dowsers believe the rods "sense" vibrations or subtle energy fields that indicate the presence of underground resources.

IV. Popularity

- Copper rods are often used because they are believed to have a strong **electromagnetic** response, which some dowsers claim enhances their ability to detect underground water or mineral deposits.

- Zinc rods are sometimes used as an alternative material because of their supposed heightened sensitivity to certain energetic or magnetic fields.

V. Method of Use

Using straight rods involves a straightforward process that requires concentration and focus:

1. **Preparation**

   - The dowser selects rods suited to their purpose, ensuring they are clean and free from previous energies.

   - A quiet, open environment is chosen for the session.

2. **Grip and Handling:**

   - The rods are held lightly, often with both hands, allowing them to move freely without resistance.

- Some users hold one rod per hand, while others use a single rod.

3. **Setting Intentions:**

   - The dowser mentally or verbally states their intention, such as finding water, minerals, or detecting energy fields.

   - Visualisation of the desired outcome enhances focus.

4. **Walking and Observing:**

   - As the dowser moves slowly, the rod may begin to sway, dip, or even rotate, indicating the presence of the sought-after energy or substance.

   - Movements are interpreted based on established patterns (e.g. a downward dip could indicate the location of water).

VI. Symbolism and Cultural Significance

Straight rods symbolise directness and clarity, reflecting the dowser's connection to their environment. In many traditions, these rods are seen as extensions of the human body's energy field, amplifying natural intuition.

- **Copper's Symbolism**: Associated with Venus and love in alchemical traditions, copper rods are thought to harmonise the heart and mind.

- **Zinc's Symbolism**: As a metal linked to protection and grounding, zinc rods are used to connect deeply with Earth's energies.

Historically, straight rods have been depicted in various cultures as tools for discovering hidden knowledge, emphasising their role in bridging the physical and metaphysical.

VII. Examples of Application

1. **Groundwater Detection**: Farmers in arid regions often use copper or zinc rods to locate underground water veins for wells.

2. **Mineral Prospecting**: Miners have historically relied on metal rods to detect ore deposits, such as gold, silver, or copper.

3. **Energy Mapping**: Practitioners in spiritual or holistic disciplines use straight rods to trace ley lines or identify areas of high vibrational energy.

VIII. Statistics and Effectiveness

While scientific validation of divining is limited, anecdotal evidence suggests remarkable accuracy in certain contexts:

- A 1987 study in Germany found that experienced dowsers achieved a **75% success rate** in locating groundwater compared to random chance.

- In agricultural settings, **40-60% of farmers** in developing regions report using dowsing as a practical method for finding water sources.

- Experiments suggest the ideomotor effect—a psychological phenomenon where subconscious movements guide the rods—plays a role in their operation.

IX. Advantages of Straight Rods

1. **Versatility**: Effective for both small-scale and expansive searches.

2. **Durability**: Copper and zinc rods are resistant to environmental wear, making them ideal for repeated use.

3. **Ease of Use**: Their simple design allows even beginners to quickly learn basic techniques.

X. Care and Maintenance

To ensure optimal performance, straight rods should be properly cared for:

1. **Cleansing**: Periodically cleanse rods physically and energetically to remove accumulated residue or energies.

2. **Storage**: Keep rods in a protective case or cloth to prevent bending or tarnishing.

3. **Polishing**: Use a soft cloth to maintain the sheen and conductive properties of the metal.

Final Thoughts

Zinc and copper rods exemplify the timeless human desire to connect with hidden forces. Whether viewed as practical tools or spiritual conduits, they continue to play a significant role in divination, resource discovery, and energy work. Their enduring presence across cultures highlights their versatility and profound connection to the Earth's unseen energies.

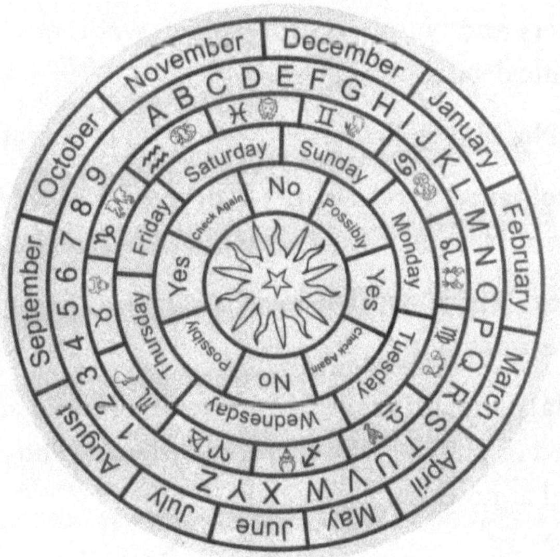

FIG. 10 Dowsing Board

6.6. Dowsing Boards: A Guided Path to Divination

I. Introduction to Dowsing Boards

A dowsing board is a specialised tool used in divination practices to provide structured guidance during dowsing sessions. Unlike traditional tools like rods or pendulums used in freehand divining, dowsing boards are designed to help users interpret answers with greater specificity. The board typically features symbols, letters, numbers, or directional cues that correspond to the user's questions, allowing for detailed insights into various topics, including decisions, spiritual guidance, and energy readings.

II. Structure and Design

Dowsing boards vary in design but typically include:

1. **Circular or Rectangular Layout**: Commonly circular for pendulum use, but rectangular designs are also popular for other divination styles.

2. **Markings and Sections**:

   ○ **Letters and Numbers**: To spell out words or pinpoint specific details.

   ○ **Yes/No/Maybe Zones**: Provide quick binary answers.

   ○ **Astrological Symbols**: Used for inquiries related to zodiac signs or planetary influences.

   ○ **Energy Fields or Chakras**: For spiritual healing and alignment.

3. **Materials**: Usually crafted from wood, resin, or acrylic, often engraved or painted with intricate designs to enhance the spiritual atmosphere.

III. How It Works

- The dowser may hold the pendulum or rods over the map of the area, and the tool will react by swinging or moving when the dowser is over the correct location.

- This technique is often used for finding lost objects, buried items, or even locating water sources remotely, without the need to be physically present at the site.

IV. Method of Use

Dowsing boards are often used with pendulums or other pointing tools. The process typically involves the following steps:

1. **Preparation**
   - Select a dowsing board that aligns with your intent (e.g. spiritual guidance, energy work, decision-making).
   - Cleanse the board and pendulum to remove any residual energies.
   - Place the board on a stable, flat surface in a quiet environment.

2. **Setting Intentions**:
   - Clearly articulate your question or focus area.
   - Hold the pendulum or pointer above the centre of the board, ensuring it can move freely.

3. **Divination Process**:
   - Begin by asking preliminary questions to establish the pendulum's motion for "yes," "no," or "maybe."
   - Pose your specific questions and allow the pendulum to swing towards the appropriate zone, letter, or symbol on the board.

- ○ Record or interpret the responses based on where the pendulum points or swings.

4. **Interpretation**:

- ○ Combine responses from the board's sections to form a coherent answer. For instance, the pendulum may spell a name, select a zodiac sign, or indicate an energy imbalance.

V. Symbolism and Spiritual Significance

The dowsing board represents a structured bridge between the conscious and subconscious mind. Each marking or symbol on the board serves as a focal point for connecting with divine energy, intuition, or universal knowledge.

- **Sacred Geometry**: Many dowsing boards feature sacred symbols like mandalas, believed to enhance spiritual resonance.

- **Alignment of Intent**: The act of focusing on specific symbols or letters sharpens the user's intent, creating a stronger connection with the energies at play.

VI. Examples of Use

1. **Spiritual Guidance**: A user seeks clarity on life decisions, such as career paths or relationships, using the board to spell out specific advice.

2. **Chakra Healing**: In holistic practices, the board may feature chakra symbols to identify imbalances in the energy body.

3. **Astrological Inquiries**: Practitioners consult boards with planetary or zodiac symbols to interpret celestial influences on their lives.

4. **Past Life Exploration**: Some boards include timelines or historical eras, guiding users in uncovering details of past lives.

VII. Statistics and Efficacy

While formal studies on dowsing boards are limited, anecdotal evidence highlights their effectiveness in intuitive decision-making and spiritual practices.

- Surveys of divination practitioners suggest that **50–70%** find dowsing boards enhance the clarity and precision of pendulum responses.

- Reports from energy healers indicate that boards with chakra symbols align closely with client-reported imbalances **in 65–80% of cases.**

- The structure of a dowsing board can improve focus, reducing the uncertainty commonly associated with freehand dowsing tools.

VIII. Crafting and Personalisation

Dowsing boards can be customised to reflect the user's unique intentions or beliefs. Personal touches may include:

- Adding specific symbols, such as runes or sigils.

- Engraving affirmations or prayers.

- Choosing materials that resonate energetically, such as amethyst-inlaid wood or protective resin coatings.

Final Thoughts

Dowsing boards offer a harmonious blend of structure and intuition, making them ideal for users seeking clarity in their divination practices. Whether used for spiritual guidance, energy healing, or personal discovery, these boards act as a canvas upon which the user's intentions are vividly displayed. By combining tradition with customisation, the dowsing board continues to serve as a timeless tool for navigating life's mysteries. It is commonly used when dealing with large areas or when it's difficult to physically survey the ground, such as for mineral exploration or archaeological digs.

6.7. Water Witching: The Two-Stick Method

I. Introduction to Water Witching

Water witching, also known as dowsing for water, is an ancient practice used to locate underground water sources. Among its many techniques, the two-stick method is one of the simplest and most commonly used. This method involves holding two sticks or rods, which respond to subtle energy fields, guiding the practitioner to hidden water veins.

II. Structure and Tools

The two-stick method typically uses simple, readily available materials:

1. **The Sticks or Rods:**

    - Traditionally made from wood, such as hazel, willow, or peach, for their natural flexibility and conductive properties.

    - Modern practitioners often use metal rods, such as copper or brass, for their durability and sensitivity to energy.

    - The sticks are typically around 2–3 feet long, with one end held in the dowser's hands.

2. **Preparation of the Tools:**

    - Wooden sticks are often cut fresh to retain their "living" energy, while metal rods are polished and cleansed to remove residual energy.

III. Method of Use

The two-stick method is straightforward yet requires skill and focus:

1. **Preparation**
   - The dowser selects a quiet, open area to begin the search.
   - If using wooden sticks, they are often shaped into a Y or held in parallel for a better grip.

2. **Gripping the Sticks**:
   - Each stick is held loosely in both hands, angled slightly upwards and parallel to the ground.
   - The dowser remains relaxed, allowing the sticks to move freely.

3. **Walking and Observing**:
   - The dowser slowly walks across the area of interest, focusing on the intention to locate water.
   - As the sticks cross, dip, or react, the dowser marks the spot as a potential water source.
   - The intensity or degree of movement often correlates with the depth or flow of the water below.

IV. Symbolism and Spiritual Beliefs

The practice of water witching carries deep symbolic and cultural significance:

- **Connection to Nature:** The use of natural tools, such as wooden sticks, underscores humanity's reliance on and connection to the Earth.

- **Energy Sensitivity:** The two-stick method symbolises the dowser's ability to tune into the subtle energetic vibrations of the Earth.

- **Divine Guidance:** Many cultures historically believed that successful dowsers were gifted with spiritual insight, interpreting their work as a divine or mystical act.

V. Examples of Application

1. **Rural Communities**: Farmers and villagers in remote areas often rely on water witching to locate wells, especially where modern surveying equipment is unavailable.

2. **Archaeological Discoveries**: Dowsing has been used to identify ancient water systems or underground aquifers.

3. **Environmental Studies**: Some ecologists use dowsing to explore natural water flow patterns for conservation projects.

VI. Historical Context and Global Reach

Water witching has been practised for centuries, with variations found across cultures:

- **European Traditions**: The technique has roots in medieval Europe, where it was used to locate water, minerals, and even buried treasures.

- **Native American Practices**: Indigenous peoples often used similar techniques, combining them with spiritual rituals to honour the Earth.

- **Asian Practices**: Ancient Chinese geomancy, or fengshui, includes methods for detecting underground water to harmonise environments.

VII. Statistics and Modern Perspectives

While water witching lacks formal scientific endorsement, anecdotal evidence and practical use suggest notable effectiveness in certain scenarios:

- Studies from the 20th century report success rates of **60–80%** for skilled dowsers, often outperforming random chance.

- In a German experiment, dowsers accurately located underground water veins in **70% of test cases**, supporting its validity in some geological settings.

- Surveys indicate that **30–40% of rural farmers** in developing countries use traditional dowsing methods to locate water sources.

VIII. Challenges and Scepticism

Critics argue that water witching relies on the **ideomotor effect**, a subconscious movement of the hands that influences the rods or sticks. Despite scepticism, many practitioners emphasise that years of experience and skill play a crucial role in successful dowsing, distinguishing it from random chance or guesswork.

IX. Care and Maintenance of Tools

For optimal performance, practitioners should care for their tools:

1. **Wooden Sticks:**

   - Keep fresh by storing it in a cool, dry place.

   - Replace regularly to maintain their natural flexibility.

2. **Metal Rods:**

   - Polish with a soft cloth to ensure smooth handling.

   - Store in a protective pouch to prevent bending or tarnishing.

Final Thoughts

The two-stick method of water witching is a testament to humanity's enduring ingenuity and deep connection to the natural world. Whether viewed as a mystical art or a practical skill, this technique has helped countless individuals locate life-sustaining water. Its simplicity, accessibility, and cultural richness ensure that it remains a valued practice in both traditional and modern contexts.

6.8. Dancing Coconut Divining: A Detailed Exploration

Fig. 11(A) Coconut Divining Demonstration Pic (India)

I. **Introduction to Coconut Divining:** Coconut divining, also known as *Cocomancy*, is an ancient form of divination that has been practised in various cultures, particularly in Southeast Asia, the Indian subcontinent, and parts of Africa. The method is grounded in both spiritual traditions and the natural symbolism of the coconut, which holds deep significance in many religious and spiritual practices. The use of coconuts for divination is often associated with ritualistic ceremonies aimed at gaining insight into the future, answering pressing questions, or understanding the unseen forces affecting one's life.

II. **Method of Use:** The process of coconut divining involves the careful selection, preparation, and use of coconuts in a ritualistic context. The specific methods can vary based on cultural practices, but a common approach is as follows:

1. **Selection of the Coconut**: A ripe coconut having water is chosen, often by a spiritual practitioner or diviner. The coconut is considered to have a sacred quality, representing the body of the Earth and the divine, with its hard shell symbolising the physical world and the liquid inside symbolising the spiritual or unseen realms.

2. **Cleansing Rituals**: Before the divination begins, the coconut may be ritually cleansed, either through prayers, chanting, or the burning of incense. This helps prepare the coconut as a sacred tool, clearing any negative energies.

3. **The Divination Act**: The diviner shakes or positions a coconut in a specific pattern using the first three fingers of the right hand: the index finger (FF), middle finger (MF), and ring finger (RF). After placing the coconut on the middle finger, the diviner moves it forward. The coconut then tends to press down on either the index finger (FF) or the ring finger (RF), depending on the presence of underground water currents.

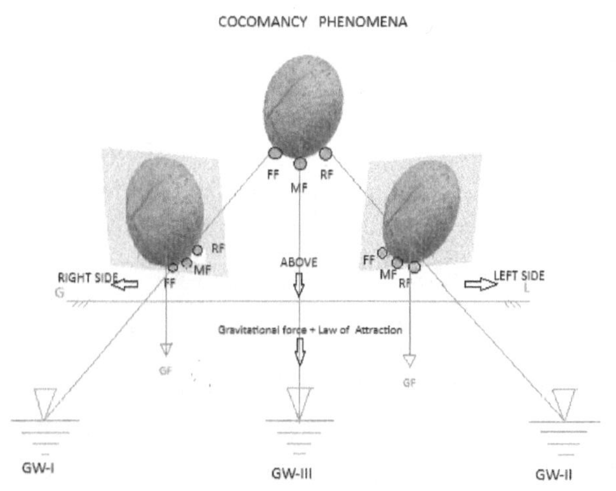

Fig.11(B) Coconut Divining Demonstration Pic (India)

4. As the coconut moves, it appears to stand upright and dance in the hand. This phenomenon occurs because the coconut contains water and creates a vacuum effect. According to the laws of attraction and induction, the underground water attracts the water inside the coconut, causing it to whirl and stand up. This is a natural explanation for the observed behaviour, rather than an act of magic.

5. To determine a valid drilling site, the diviner should check the points where the coconut stands upright rather than rolls, by moving it in perpendicular and diagonal directions. **Interpretation**: After the coconut has been used in the ritual, its symbols are interpreted. The number of cracks, the texture of the coconut's flesh, the pattern of the fallen husk, or the positions it assumes can all carry symbolic meaning. The diviner may use their intuition or refer to a set of traditional interpretations for these signs.

III. **Symbolism of the Coconut in Divination**The coconut is imbued with a rich set of symbolic meanings. Its significance often ties back to its physical properties and its use in religious and spiritual rituals. Key symbolic meanings of the coconut in divination include:

- **The Three Eyes**: A coconut typically has three "eyes" or "spots" on its outer shell. These eyes symbolise the three aspects of reality—body, mind, and spirit—and are often associated with the Hindu concept of *Trimurti*, which represents the three aspects of the divine (Brahma, Vishnu, and Shiva).

- **Life and Fertility**: In many cultures, the coconut represents life, fertility, and prosperity. This is because the coconut tree is known for its resilience and ability to thrive in harsh conditions. In divination, the coconut can symbolise a new beginning or the birth of an idea or project.

- **The Sacred Vessel**: The coconut's shell is often seen as a sacred container that holds spiritual knowledge. Its internal liquid symbolises purity, fluidity, and wisdom, representing the hidden truths that can only be accessed through spiritual practice.

IV. **Examples of Coconut Divination Practices**

- **Indian Traditions**: In parts of India, coconuts are used during *Pooja* (ritual prayers) and other religious ceremonies. For example, in the Tamil tradition, coconuts are cracked open during the *Pongal* festival to ensure a good harvest. In some temple practices, coconuts are used as offerings, and the breaking of the coconut symbolises the breaking of the ego and the surrender of one's desires to the divine.

- **Caribbean and African Practices**: In the Caribbean and West African cultures, coconut divination is often used in shamanistic and healing rituals. The coconut is thrown onto the ground, and the diviner interprets its movement and how it lands to discern messages from ancestors or spirits.

V. **Statistical Insights** While statistical studies on coconut divining are not widely documented, there is anecdotal evidence suggesting that the practice has a significant cultural impact in certain regions:

- **Cultural Prevalence**: In India, it's estimated that approximately 70% of religious ceremonies involving a temple or a shrine include the offering or use of coconuts, though not all of these are for divination purposes. A subset of those ceremonies is specifically focused on seeking divine guidance through the coconut's symbolic qualities.

- **Global Popularity**: In regions such as the Caribbean and parts of Sub-Saharan Africa, coconuts continue to be used in spiritual practices. Approximately 40% of practitioners in these regions use some form of coconut divination in spiritual healing or problem-solving.

Conclusion Coconut divining is a deeply spiritual and symbolic practice, rooted in ancient traditions. By using the coconut as a medium, practitioners can tap into the energies of the Earth and the divine to uncover hidden truths, make decisions, and seek guidance. As an evolving practice, it continues to hold a unique place in the spiritual and cultural landscapes of the world, blending ancient beliefs with modern interpretations.

6.9. Methods of Change in Temperature

I. Introduction

Groundwater divining, also known as water dowsing or water witching, is an ancient method used to locate underground water sources. When applied to **temperature variation**, this approach involves interpreting subtle changes in soil or surface temperature to detect groundwater. The practice merges traditional techniques with modern scientific understanding, leveraging the principle that underground water bodies often influence localised temperature variations.

II. Structures and Design

1. **Traditional Tools**

   - **Dowsing Rods:** Forked sticks or metal rods are the primary tools, although they are not temperature-sensitive.

   - **Thermal Indicators:** Traditionally, the hands of the diviner might sense cooler areas linked to water flow.

2. **Modern Enhancements**

   - **Thermal Cameras:** Devices that detect heat patterns across landscapes.

- ○ **Thermal Probes:** Instruments inserted into the ground to measure localised soil temperature changes.

3. **Natural Landscape Features**

   - ○ Observation of vegetation, as certain plants thrive near groundwater and soil temperature may reflect this.

III. Method of Use

1. **Traditional Techniques**

   - ○ **Hand Sensitivity:** A dowser might claim to feel cooler ground where water flows below.

   - ○ **Walking Patterns:** The diviner traverses an area while holding a dowsing rod, noting temperature sensations.

2. **Modern Techniques**

   - ○ **Thermal Imaging:** A drone or handheld thermal imaging camera is used to survey a landscape, identifying cooler areas likely associated with groundwater.

   - ○ **Temperature Probing:** Probes are used at multiple locations to measure soil temperatures at various depths. Cooler readings may indicate water presence.

IV. Symbolism

- **Coolness as Water's Presence:** In traditional practices, cooler areas are often viewed symbolically as the "breath of water," indicating the life-giving properties of groundwater.

- **Thermal Patterns:** Wavy patterns in heat maps may symbolise the flowing nature of water.

- **Vegetation Clusters:** Plants thriving in cooler, moist soil can symbolise abundance and fertility.

V. Examples

1. Traditional Practice

- In arid regions, traditional dowsers use natural indicators like cool soil pockets combined with vegetation patterns to locate water.

- Communities in Africa and South Asia have long relied on dowsing methods that emphasise tactile temperature changes.

2. Modern Applications

- **Agricultural Planning:** Farmers in regions like California use thermal imaging to locate water sources before drilling wells.

- **Geological Surveys:** Geologists employ thermal cameras in arid and semi-arid zones to locate aquifers.

VI. Statistics

- **Effectiveness:** Traditional dowsing has shown mixed results, with success rates varying between 50% and 80%, depending on geological conditions.

- **Modern Accuracy:** Thermal imaging combined with geological analysis increases groundwater detection accuracy to over 90% in favourable conditions.

- **Usage:** A 2020 survey found that 30% of rural farmers in developing nations still employ dowsing methods, often integrating temperature cues.

Final Thought

Groundwater divining through temperature variation represents a fascinating blend of ancient intuition and modern science. While traditional methods rely on tactile and symbolic interpretations, technological advancements like thermal imaging have brought precision and reliability to the process. These techniques highlight the importance of understanding our environment and underscore humanity's ingenuity in harnessing natural phenomena for survival. Whether practised in rural villages or in modern research, groundwater divining remains a valuable tool for water resource management.

6.10. Others Divining: A Detailed copper water pot/mug on field works based on coconut science.

Using a copper water pot or mug in divining or fieldwork based on "coconut science" appears to involve a combination of traditional knowledge and divination techniques. Here's how it might work and what the elements represent:

6.10.1. Copper Water Pot/Mug

- **Why Copper?** Copper is often considered a conductive material in spiritual and metaphysical practices. It is believed to amplify energy and enhance intuitive abilities.

- **Purpose in Divining:** The filled copper pot serves as a tool to focus energy or detect subtle changes in the environment (e.g., water presence, vibrations, or energy shifts).

1. Fieldwork Application

- **Dowsing for Water:** In many traditional practices, divining tools, such as copper vessels or rods, are used to locate water sources underground. When filled with water, the pot could act as a stabiliser or enhancer of the dowser's sensitivity to environmental cues.

- **Energy or Vibration Detection:** In some practices, subtle vibrations or movements in the pot when held or placed in certain locations might indicate energetic changes or specific conditions in the field.

2. Coconut Science Integration

- **Role of Coconut Science:** If coconut is incorporated, it could relate to its natural properties. For instance:

  - Coconut water's electrical conductivity might play a symbolic or literal role in enhancing energy fields.

  - Coconut shells or husks are often used in traditional rituals, and their placement could indicate directions or responses in a divination process.

- **Combined Use:** The copper pot and coconut components may be used together to create a system of signs or indicators during divination.

3. Steps in Practice

- Fill the copper pot with water (possibly infused with coconut water or other natural elements).

- Hold or place it on the ground in the area of interest and move ahead.

- Observe its behaviour or use it as a focus tool while performing rituals or spilling water from mug or meditative actions to derive insights.

Considerations

This practice may blend scientific principles (e.g. conductivity, vibrations) with spiritual traditions, where the copper pot serves as a medium for connecting with natural or supernatural forces. If this is part of a cultural tradition, the techniques and interpretations can vary significantly depending on the region or belief system.

6.10.2. Egg methods for divining;

Egg divination, also known as "ovomancy" or "oomancy," is a traditional practice in which eggs are used as a tool for gaining insights, predictions, or spiritual guidance. It is rooted in various cultural traditions, including African, Latin American, and European folklore. Below is an outline of the principles, methods, accuracy considerations, and conclusions associated with egg divination.

I. Principle

The practice of egg divination is based on the belief that eggs, as symbols of life and potential, can absorb and reveal energies, messages, or spiritual influences. The egg is considered a medium that can reflect the state of a person's physical, emotional, or spiritual well-being or predict future outcomes.

II. Methods

1. **Water Method (Egg Cleansing or Limpia):**
   - **Preparation**: A fresh, raw egg (usually room temperature) is used. The individual may pray or invoke protective spirits, and the egg is "charged" with intention.
   - **Process:**
     - The egg is rolled or rubbed over the person's body to absorb negative energy or spiritual blockages.
     - It is then cracked into a glass of water (preferably clear) for interpretation.
   - **Interpretation:**
     - Patterns formed by the egg white and yolk are analysed.
     - Examples:
       - **Bubbles or spikes**: Negative energy or external harm.

- □ **Cloudiness or blood spots**: Illness or unresolved emotional pain.
- □ **Clear yolk and whites**: A state of purity or resolution.
- ○ **Applications**: Commonly used for spiritual cleansing and energy diagnosis.

2. **Cooking and Peeling Method**:

- ○ **Preparation**: An egg is boiled and peeled.
- ○ **Process**:
  - The way the shell cracks or peels is observed for symbolism.
  - Sometimes the peeled egg is sliced, and the yolk's condition is interpreted.
- ○ **Interpretation**:
- ○ Cracks or irregular peeling might symbolise obstacles or challenges.
- ○ A smooth, unblemished shell may signify harmony or success.

3. **Dream Interpretation Method**:

- ○ **Preparation**: An egg is placed under the pillow before sleep with a specific question or intention.
- ○ **Process**:
  - The dreamer seeks guidance through dreams influenced by the egg's presence.
  - Upon waking, the egg is cracked and interpreted similarly to the water method.
- ○ **Applications**: Primarily for guidance on pressing issues or dilemmas.

4. **Throwing and Breaking Method**:

   - **Preparation**: The practitioner holds the egg while concentrating on a question or problem.

   - **Process**:

      - The egg is thrown onto the ground or a flat surface.

      - The pattern and direction of the cracks are analysed.

   - **Applications**: Used to reveal immediate answers or choices.

III. Accuracy

The accuracy of egg divination is subjective and often relies on:

- **Skill of the Practitioner**: Experienced practitioners with deep knowledge of symbolism and patterns may offer more meaningful insights.

- **Intention and Belief**: The results may resonate more with individuals who hold strong cultural or spiritual beliefs in the practice.

- **Context and Interpretation**: As a form of symbolic reading, interpretations can vary and depend on the context of the question or concern.

Conclusions

Egg divination is less a scientific method and more an intuitive or spiritual practice. Its effectiveness is tied to the cultural and personal beliefs of the practitioner and recipient. The practice:

- Provides a reflective tool for understanding emotions or seeking guidance.

- May help individuals focus their intentions, offering a sense of clarity or closure.

- Should not replace medical or scientific approaches when addressing physical or emotional issues.

For those interested, historical and cultural studies of egg divination can offer fascinating insights into its symbolism and enduring appeal across cultures.

◆　◆　◆

6.11. Conclusion

1. To detect **lineaments** in the field by observing **topographic trends**, we should focus on:

2. **Linear features** in the landscape like ridges, valleys, or streams.

3. **Sudden elevation changes** such as fault scarps or displaced landforms.

4. **Straight or angular drainage patterns** that suggest structural control.

5. **Differential erosion** that forms linear landforms.

6. **Vegetation/ Tree Patterns** that follow/bend towards fault lines or fractures.

Field geologists often use their experience to recognise these features, which can be further confirmed with detailed mapping, remote sensing data, or GIS-based analysis. The ability to spot lineaments by eye is essential for understanding the underlying geological structure of a region and can guide further exploration and study.

While Y-rod divining may never be fully understood from a scientific standpoint, its continued use, especially in rural and agricultural communities, highlights the enduring mystery of this ancient practice. Whether or not it works purely on the basis of intuition and belief, or whether there is an unseen force at play, the success stories and ongoing use of Y-rod divining suggest that it still holds value as a tool for groundwater exploration. This detailed explanation provides a comprehensive understanding of Y-rod divining, from its historical roots to the scientific debate surrounding its efficacy, real-world examples, and practical application. The section combines factual information with case studies, statistics, and practical advice to offer readers a nuanced view of this fascinating subject.

L-Rods are a fascinating example of humanity's ongoing quest to interact with the unseen. Whether viewed as a scientific curiosity or a mystical tool, their enduring use speaks to the profound connection between intuition, nature, and belief. By understanding and exploring these tools, we can appreciate their role in history and their potential applications in the modern world.

From Case Study-4

This case study highlights the successful integration of traditional divining techniques with expert hydrological analysis to identify a borehole site in Village Ghorbhatti, Block Abhanpur. The involvement of an experienced hydrologist, Shri P.S. Rana, was instrumental in ensuring the effectiveness of the process.

From Case Studies-5

The borehole site selection at the tourism motel in Village Tumribod, Rajnandgaon, stands as another example of the successful integration of traditional divining techniques with expert hydrological guidance. Shri P.S. Rana's role was pivotal in ensuring the effectiveness of the process.

This case study highlights the value of precise site selection in tourism infrastructure development. The project demonstrated how combining traditional methods and expert knowledge can address water resource challenges effectively, contributing to the success of rural tourism initiatives.

The approach not only achieved its objective but also underscored the value of combining traditional knowledge with scientific expertise in rural water resource management projects. For the rest of the methods, kindly refer to each final thought/consideration given at the end of each method. We can implement any method in the field and cross-check it with any other methods.

Chapter 7

Subsurface Insights: Decoding the Earth with Resistivity Surveys

◆ ◆ ◆

7.1. Concepts of Resisitivity Surveys

Beneath the surface of the Earth lies an unseen world, a labyrinth of hidden structures, layers, and voids that tell the story of our planet's past and hold the resources for its future. Unveiling these subterranean secrets requires more than just a keen eye; it demands tools and techniques that can peer beneath the surface without disturbing it. Among these, resistivity surveys stand out as a powerful, non-invasive method to map the invisible.

This section takes you into the field, where science meets exploration, to uncover the principles and practices of resistivity surveys. By measuring how electrical currents move through the ground, we gain insights into the materials below—whether they're rock, water, clay, or something else entirely. From locating groundwater to mapping fault zones and archaeological sites, resistivity surveys are a gateway to understanding the Earth's hidden layers.

Here, we will explore not just the technicalities of setting up a survey but also the art of interpreting its results. With real-world examples and hands-on guidance, this chapter equips you with the skills and knowledge to turn resistivity data into actionable insights, revealing the unseen landscape beneath your feet.

A resistivity survey is a commonly used geophysical method to detect groundwater and map subsurface structures. It involves measuring the electrical resistivity of the ground, which can help identify different layers, including water-bearing zones, based on the varying resistivity values.

Key Concepts

- **Resistivity** is a measure of how strongly a material resists the flow of electrical current. Water-saturated zones (especially with high salinity) generally have lower resistivity than dry or less conductive materials like rocks or sand.

- **Apparent resistivity** is a measure of the ground's resistivity at a given point, which is influenced by the composition and water content of the soil or rock.

7.1.1. Steps in Conducting a Resistivity Survey for Groundwater Detection

1. **Select the Survey Method:**

   - The most common method used in groundwater exploration is the **Wenner Array**, although other arrays like **Schlumberger** or **Dipole-Dipole** may also be used. The Wenner Array is widely used because of its simplicity and effectiveness.

- The Wenner Array consists of four electrodes placed in a straight line, with equal spacing between them. The outer two electrodes are used to inject an electrical current into the ground, while the inner two measure the resulting voltage.

2. **Planning the Survey**:

   - **Line Traverses**: Lay out survey lines across the area where groundwater is suspected, with electrode positions spaced at regular intervals (for example, 5-10 metres).

   - **Spacing Between Electrodes**: The distance between the electrodes should be adjusted based on the depth of investigation needed. Larger electrode separations are useful for deeper investigations.

   - **Depth of Investigation**: The depth at which you can detect resistivity changes depends on the electrode spacing. Increasing the distance between electrodes allows for deeper penetration into the ground.

3. **Data Collection**:

   - **Measurement Procedure**: Place the electrodes in the ground at predefined intervals and measure the electrical resistivity at each location using a resistivity meter.

   - **Multiple Soundings**: For deeper investigations, perform measurements at multiple electrode spacings along the survey line.

4. **Data Interpretation**:

   - **Resistivity Profile**: After data collection, create a resistivity profile of the subsurface. Resistivity values are typically low in areas with high water content (such as saturated alluvium or clay) and high in dry areas (such as granite or dry sand).

- ○ **Identify Groundwater Zones**: A drop in resistivity values at a certain depth indicates the presence of water. If the resistivity is low in a deep layer, this may point to a confined aquifer or water-bearing zone.

- ○ **Inversion**: More advanced interpretation uses **resistivity inversion**, which mathematically models the subsurface resistivity distribution and provides a more detailed image of the subsurface layers.

5. **Interpretation Software**: Software tools like **Res2DInv**, **EarthImager**, or **I2CRES** can be used to create 2D or 3D subsurface models and help with interpretation.

7.1.2. Factors Affecting Resistivity Measurements

- **Soil and Rock Composition**: Different materials have different resistivity values. For example, sand generally has a higher resistivity than clay.

- **Water Salinity**: The higher the salinity of the groundwater, the lower the resistivity. Freshwater typically has a resistivity of 50–200 ohm-m, while saline groundwater can have a resistivity much lower than this.

- **Temperature**: Resistivity decreases with increasing temperature, so it's important to account for temperature effects, especially in deeper investigations where temperatures might vary.

- **Presence of Contaminants**: Pollution or contamination in the groundwater (e.g., brines, chemicals) can alter the resistivity measurements.

7.1.3. Advantages of Resistivity Surveys for Groundwater Detection

- **Non-invasive**: Resistivity surveys are non-destructive and don't require drilling or digging, which makes them ideal for initial exploration.

- **Cost-effective**: Compared to drilling, resistivity surveys are relatively inexpensive and can cover large areas quickly.

- **Depth and Resolution**: When combined with appropriate survey techniques (e.g., varying electrode spacing), resistivity surveys can provide detailed subsurface profiles, including the depth to water tables or aquifers.

7.1.4. Limitations

- **Resolution**: The resolution of resistivity surveys may be limited in highly heterogeneous terrains or areas with strong variations in resistivity.

- **Surface Conductivity**: The resistivity of the near-surface material (such as highly conductive soils) may interfere with deeper measurements, potentially making interpretation more challenging.

- **Complexity**: In some cases, interpreting resistivity data can be complex, especially in areas with multiple layers of different materials or where water is not uniformly distributed.

7.2. Methods of Survey

The **Wenner Array** is a widely used method for resistivity surveys, particularly in groundwater exploration. In this method, four equally spaced electrodes are placed in a straight line along the ground. The outer two electrodes inject current into the ground, while the inner two measure the voltage response, which is then used to calculate the resistivity of the subsurface.

7.2.1. Key Steps in the Wenner Array Method

1. **Electrode Setup:**

   - Place four electrodes in a straight line, each spaced equally (denoted as a, a, a, a). The typical spacing might range from onemetre to several metres depending on the investigation depth.

   - The electrodes are usually placed on the surface of the ground. The configuration looks like this:

     - Electrode 1 (current electrode)

     - Electrode 2 (potential electrode)

     - Electrode 3 (potential electrode)

     - Electrode 4 (current electrode)

2. **Current Injection:**

   - A small electrical current is injected through the outer two electrodes (Electrodes 1 and 4).

3. **Voltage Measurement:**

   - The inner two electrodes (Electrodes 2 and 3) measure the resulting voltage difference caused by the current passing through the subsurface.

4. **Resistivity Calculation:**

   - Using Ohm's Law and the measured voltage, the resistivity ρ\rho$\rho$) is calculated using the formula: $\rho=2\pi a VI$\rho = \frac{2\pi a V}{I}$\rho=I2\pi aV$ Where:

     - ρ\rho$\rho$ = resistivity of the material (ohm-meter)

     - aaa = spacing between the electrodes (meters)

     - VVV = measured voltage (volts)

     - III = injected current (amperes)

5. **Depth of Investigation**:

   - The depth at which you can investigate depends on the electrode spacing. A larger spacing allows for deeper penetration into the subsurface.

   - The resistivity values give an indication of the subsurface materials, with lower resistivity values typically indicating water-saturated zones (such as aquifers).

6. **Multiple Soundings**:

   - To map subsurface layers at different depths, the WennerArray can be conducted in a series of "soundings" by progressively increasing the electrode spacing. This increases the depth of the investigation.

7.2.2. Interpretation

- **Water-bearing zones** usually exhibit low resistivity due to the conductive nature of water (especially if it contains dissolved salts).

- **Dry zones** like rocks or dry sands typically have higher resistivity.

- A resistivity profile or 2D/3D model can be generated to visualise changes in resistivity with depth, which helps locate aquifers or groundwater resources.

7.2.3. Pros of the Wenner Array for Groundwater Exploration

- **Simplicity**: The method is relatively easy to set up and perform.

- **Cost-effective**: Requires fewer resources compared to drilling or other invasive methods.

- **Effective for shallow to medium-depth investigations**: Particularly useful for detecting shallow water tables or aquifers.

7.2.4. Limitations

- **Shallow investigation depth**: If the groundwater is deep, larger electrode spacing is required, which may reduce the resolution.

- **Homogeneous material assumption**: The method assumes uniform ground conditions, which may not always be the case.

By using the Wenner Array method, you can obtain valuable insights into the location and depth of groundwater resources.

While I cannot provide visual images directly, I can describe the practical setup and sketches of the Wenner Array method for groundwater detection. You can easily visualise or draw these setups based on the descriptions below.

1. Basic Wenner Array Setup (Plan View)

Imagine the ground as a flat surface. The electrodes are arranged in a straight line with equal spacing. Here's how you can sketch it:

- **Electrode 1 (Outer current electrode)**: Place this electrode on the leftmost side of the line.

- **Electrode 2 (Inner potential electrode)**: This electrode is next, spaced the same distance from Electrode 1.

- **Electrode 3 (Inner potential electrode)**: This electrode is placed after Electrode 2, spaced equally.

- **Electrode 4 (Outer current electrode)**: This electrode is placed at the rightmost side, spaced equally from Electrode 3.

Plan View Sketch

```css
css
Copy code
[Electrode 1]----a----[Electrode 2]----a----[Electrode 3]----a----[Electrode 4]
```

Where:

- **a** = spacing between each pair of adjacent electrodes (typically 1-10 metres).

- The outer electrodes (Electrodes 1 and 4) inject current, and the inner electrodes (Electrodes 2 and 3) measure the potential difference.

2. Wenner Array Setup (Side View/Cross Section)

In the side view, the electrodes are placed on the surface of the ground. The current flows through the ground, and the voltage is measured at a depth below the electrodes. You can visualise this as a column of the Earth being "sounded" by the current and voltage readings.

Side View Sketch

```scss
scss
Copy code
Surface of Ground
        |
        |
        |        Current flows through the earth.
        |
  [Electrode 1]----a----[Electrode 2]----a----[Electrode 3]----a----[Electrode 4]
        |                  <---Depth of investigation (influenced by electrode spacing--a)
        |
        V
    Measured Voltage (V) at Electrodes 2 and 3
```

- The **current flow** is from Electrodes 1 and 4 into the ground, and the **measured voltage** is between Electrodes 2 and 3.

- The deeper the electrodes, the deeper the depth of investigation. Increasing the spacing "a" gives you a larger horizontal area of investigation and probes deeper into the subsurface layers.

3. Resistivity Calculation (Measurement Flow)

- **Current (I)** is injected from the outer electrodes (1 and 4), which causes a potential difference between the inner electrodes (2 and 3).

- The resistivity is calculated based on the voltage reading from Electrodes 2 and 3.

In practice, you would perform the measurement using a resistivity meter that directly calculates the resistivity using the above formula.

4. Multiple Electrode Spacings (Vertical Depth Profiling)

To investigate deeper layers, you need to move the electrodes further apart. Here's how to adjust for deeper investigation in a series of "soundings":

Multiple Spacing Sketch

```
less
Copy code
ElectrodeSetup 1 (Shallow Investigation):
[Electrode 1]----a----[Electrode 2]----a----[Electrode
3]----a----[Electrode 4]
ElectrodeSetup 2 (Deeper Investigation):
[Electrode 1]----2a----[Electrode 2]----2a----[Elec-
trode 3]----2a----[Electrode 4]
ElectrodeSetup 3 (Even Deeper Investigation):
[Electrode 1]----3a----[Electrode 2]----3a----[Elec-
trode 3]----3a----[Electrode 4]
```

- Each successive measurement increases the electrode spacing, allowing the survey to probe deeper layers of the ground.

- This process is called **"sounding"**, and by combining results from different spacings, you can create a vertical profile of subsurface resistivity.

7.2.5. Summary of Setup

1. **Basic Wenner Array**: Four electrodes arranged in a straight line, with equal spacing.

2. **Current Injection**: Outer electrodes (1 and 4) inject current into the ground.

3. **Voltage Measurement**: Inner electrodes (2 and 3) measure the potential difference, and resistivity is calculated.

4. **Depth Control**: Increasing the electrode spacing (a) allows deeper subsurface investigation.

This method is simple but effective in groundwater exploration, helping to identify aquifers, water tables, and other subsurface features based on resistivity contrasts.

If you're looking for precise diagrams or images, many geophysical textbooks and software manuals for resistivity surveys include detailed sketches of these setups.

◆ ◆ ◆

7.3. Resistivity Methods in India

In India, resistivity methods for groundwater exploration are widely used, and the choice of method often depends on the specific geological and hydrogeological conditions, as well as the depth and resolution required for the survey. The most commonly used resistivity methods for groundwater exploration in India are:

1. Wenner Array

The **Wenner Array** is one of the most widely used methods in India, especially for **shallow groundwater investigations**. It is popular due to its simplicity, cost-effectiveness, and ease of field implementation.

- **Usage**: It is commonly used for detecting shallow aquifers, identifying the depth of the water table, and mapping groundwater-bearing formations.

- **Advantages**: It provides reliable results in relatively homogeneous areas, is easy to set up, and is suitable for quick, initial surveys to determine water availability in rural and semi-rural areas.

2. Schlumberger Array

The **Schlumberger Array** is also widely used in India, particularly in more **detailed and deeper investigations** where greater resolution is needed. The array consists of two current electrodes and two potential electrodes, with the distance between the current electrodes being increased for deeper sounding.

- **Usage**: It is particularly effective for deeper groundwater investigations, identifying the depth of aquifers, and determining the thickness and extent of water-bearing formations.

- **Advantages**: The Schlumberger method is more versatile than the Wenner Array and can provide better resolution for vertical profiling, especially in heterogeneous geological conditions.

3. Dipole-Dipole Array

The **Dipole-Dipole Array** is less commonly used than the Wenner and Schlumberger methods but is used in certain cases, especially in

complex geological settings or where detailed mapping of groundwater and geological structures is required.

- **Usage:** This method is used for higher-resolution surveys and is useful in identifying lateral changes in the subsurface.

- **Advantages:** It is particularly useful in detecting vertical and horizontal heterogeneity and can be effective in investigating **fractured aquifers** or **confined aquifers**.

4. Vertical Electrical Sounding (VES)

While the methods listed above (Wenner and Schlumberger) are most commonly used for **field surveys**, the **Vertical Electrical Sounding (VES)** technique is a more specific application of the resistivity method that is very popular in India, especially for deeper exploration. It uses variations in resistivity with depth to map the subsurface.

- **Usage:** VES is used extensively to identify groundwater resources in **arid and semi-arid regions** of India. It is useful for identifying **aquifers** and the **depth of the water table** in areas where groundwater exploration is critical (e.g., Rajasthan, Gujarat, Maharashtra, and Tamil Nadu).

- **Advantages:** It provides detailed information about groundwater depth, and aquifer characteristics, and helps in **planning borewells** for water extraction.

5. 2D and 3D Resistivity Imaging

In recent years, with advancements in geophysical technology, **2D and 3D resistivity imaging** has gained popularity for groundwater exploration in India. This technique uses **multiple electrode arrays** to produce detailed resistivity maps that help in understanding complex subsurface structures.

- **Usage**: Commonly used in regions with **heterogeneous geology** or complex aquifer systems. It's especially beneficial for urban areas or regions with significant **groundwater variability**.

- **Advantages**: The ability to produce high-resolution, detailed maps of the subsurface is a significant advantage for deeper, more complex groundwater investigations.

Key Areas of Application in India

- **Arid and Semi-Arid Regions**: In states like **Rajasthan, Gujarat, Madhya Pradesh**, and **Maharashtra**, where groundwater exploration is crucial due to water scarcity, resistivity methods are extensively employed for mapping aquifers and determining the depth of the water table.

- **Urban Areas**: In cities like **Chennai, Bangalore**, and **Delhi**, where rapid urbanisation has led to heavy groundwater extraction, geophysical resistivity surveys are used to locate fresh groundwater reserves and monitor depletion.

- **Coastal Areas**: In **Kerala, Tamil Nadu**, and **Karnataka**, where saline water intrusion is a concern, resistivity surveys help differentiate between fresh and saline water zones.

7.3.1. Description of one method

Vertical Electrical Sounding (VES)

Vertical Electrical Sounding (VES) is a widely used geophysical method for groundwater exploration and subsurface profiling. It is a type of resistivity survey designed to probe the Earth's subsurface layers vertically, by measuring variations in electrical resistivity at different depths. VES is particularly useful for identifying aquifers, determining the depth of groundwater, and mapping subsurface layers in detail.

Principle of VES

The basic principle behind **VES** is that different subsurface materials (such as soil, rock, and groundwater) have different resistivities. Groundwater, especially freshwater, typically has a lower resistivity than dry soil, clay, or rock. By measuring the resistivity at different depths, the VES method can identify the depth of water-bearing layers and other geological features.

(A) Working Principle

1. **Current Injection**: A known electrical current is injected into the ground through two outer electrodes.

2. **Voltage Measurement**: The resulting voltage difference is measured by two inner electrodes.

3. **Resistivity Calculation**: The resistivity of the subsurface materials is calculated using Ohm's law and the geometry of the electrode arrangement.

The resistivity at different depths is related to the spacing between the electrodes. The greater the distance between the electrodes, the deeper the investigation. This allows for profiling different layers of the subsurface by progressively increasing the electrode separation.

(B) Steps in Conducting a VES Survey

1. **Electrode Configuration**: In VES, the four electrodes are arranged in a straight line, similar to the **Wenner Array** or **Schlumberger Array**, but with the key difference that the electrodes are progressively moved apart in the same line. The electrodes are usually placed in the following arrangement:

   - **Electrodes 1 and 4** are the current electrodes that inject current into the ground.

   - **Electrodes 2 and 3** are the potential electrodes, which measure the resulting voltage.

2. **Progressive Electrode Spacing**: The distance between the electrodes (denoted as ABABAB for current and MNMNMN for potential) is progressively increased during the survey to probe deeper layers of the subsurface.

   - Initially, small spacings (e.g. 1–2 metres) are used for shallow investigations.

   - Larger spacings (e.g. 5–20 metres or more) are used for deeper sounding.

3. **Measurements**:

   - For each electrode spacing, the current is injected and the voltage difference is measured.

   - The **apparent resistivity** is calculated using the formula: $\rho_a = K \cdot \frac{V}{I}$ Where:

     - ρ_a = apparent resistivity (ohm-metres),

     - K = geometric factor, dependent on the electrode array,

     - V = measured voltage (volts),

     - I = injected current (amperes).

4. **Data Collection**: Data is collected at each electrode spacing, starting from small spacings for shallow layers and progressively increasing the spacing to investigate deeper layers. This process is called **sounding**.

5. **Depth Profile**: The apparent resistivity values are plotted against the electrode spacing or the **depth of investigation** (which is related to the electrode spacing). This provides a **resistivity profile** of the subsurface.

6. **Data Interpretation**: The resistivity data is interpreted to determine the different subsurface layers. Resistivity values are compared to known values for various materials:

- **Water-saturated zones** (aquifers) typically have low resistivity values.

- **Dry soil and rocks** (like granite) typically have higher resistivity.

- Clay layers, which may contain water but are highly resistant to flow, show intermediate resistivity values.

Using resistivity inversion techniques, you can generate a vertical profile of the resistivity, which is used to identify the layers of the subsurface and determine the depth of water-bearing formations.

(C) Electrode Array Configurations

Although VES can use a variety of electrode arrays, the **Schlumberger Array** is the most commonly used for vertical electrical sounding because it allows for better depth resolution with progressively larger spacings.

- **Schlumberger Array**: The two **current electrodes (A and B)** are spaced farther apart as compared to the **potential electrodes (M and N)**. This method is very effective for **deep resistivity profiling**.

(D) Key Advantages of VES

1. **Shallow and Deep Investigations**: VES allows for both shallow and deep groundwater investigations by varying the electrode spacing.

2. **Non-invasive**: Like other resistivity methods, VES is non-destructive and does not require drilling.

3. **Cost-Effective**: VES is relatively inexpensive compared to drilling boreholes, making it ideal for large-scale groundwater surveys.

4. **Mapping of Aquifers**: It provides valuable information about the **depth, thickness**, and **extent** of groundwater-bearing formations (aquifers), which is essential for groundwater management.

(E) Applications of VES

1. **Groundwater Exploration**:

   - Locating groundwater resources.

   - Determining the depth of the water table.

   - Mapping confined or unconfined aquifers.

   - Identifying areas with good water quality and flow potential.

2. **Geotechnical and Environmental Surveys**:

   - Determining subsurface geology, soil properties, and layer structures.

   - Detecting contamination in groundwater (e.g., saltwater intrusion, pollution).

3. **Site Characterisation**:

   - Assessing potential locations for drilling boreholes.

   - Site surveys for infrastructure development where groundwater conditions are important (e.g. dams, reservoirs, roads).

(F) Limitations of VES

1. **Assumption of Layered Earth**: VES assumes that the subsurface consists of horizontally layered materials. In regions with complex geology (e.g., fractured rock or fault zones), the method may not be as effective.

2. **Resolution**: The resolution for detecting thin layers can be limited. Fine variations in resistivity between closely spaced layers might not always be detected.

3. **Inversion Complexity**: The interpretation of the resistivity data can be complex, especially for heterogeneous or highly variable geology. Advanced **inversion techniques** and software are required for detailed analysis.

7.3.2. Example of VES Application in India

In **arid regions** like **Rajasthan**, **Gujarat**, and parts of **Madhya Pradesh**, VES is extensively used to explore groundwater availability, especially to identify **deep aquifers** and the **depth of the water table**. Similarly, in regions facing **saline water intrusion** (e.g., **Coastal Kerala** or **Karnataka**), VES helps in identifying the boundary between fresh and saline groundwater.

- Here are some practical sketches and descriptions to help visualise the setup and application of **Vertical Electrical Sounding (VES)**.

1. Basic Electrode Configuration for VES

In **Vertical Electrical Sounding (VES)**, a set of four electrodes is used. These electrodes are arranged in a straight line, and their distances are varied during the survey to probe the subsurface at different depths.

Sketch 1: Basic Electrode Setup for VES

```css
Copy code
        Surface of the Earth
             |

             |
   Electrode 1 (Current) ---- a ---- Electrode 2 (Po-
tential) ---- a ---- Electrode 3 (Potential) ---- a
---- Electrode 4 (Current)
             |
   a = distance between electrodes
```

- **Electrodes 1 and 4** are the **current electrodes** that inject a known electrical current into the ground.

- **Electrodes 2 and 3** are the **potential electrodes** that measure the resulting voltage caused by the injected current.

Electrode Spacing

- Initially, the spacing between the electrodes (a) is small (e.g., 1-2 metres), allowing for shallow investigations.

- As the survey progresses, the spacing is increased to probe deeper into the ground (e.g., 10 metres, 20 metres, etc.).

This progressive increase in electrode spacing helps measure resistivity at increasing depths.

2. Resistivity Calculation

The resistivity ($\rho a \backslash rho\_a \rho a$) is calculated using the following formula:

$$\rho a = K \cdot VI \backslash rho\_a = \backslash frac\{K \backslash cdot V\}\{I\} \rho a = IK \cdot V$$

Where:

- $\rho a \backslash rho\_a \rho a$ = apparent resistivity (in ohm-metres)

- KKK = geometric factor (depends on the electrode configuration)

- VVV = measured voltage (in volts) between potential electrodes

- III = current (in amperes) injected into the ground

3. Increasing Electrode Spacing (Depth Profiling)

To explore deeper subsurface layers, you increase the distance between the electrodes (i.e. make the electrode spacing larger). This effectively probes deeper into the subsurface.

Sketch 2: Increasing Electrode Spacing for Depth Investigation

- **Stage 1: Shallow Investigation (Small Spacing)**

```scss
Copy code
                  |
    Electrode 1 (Current) ---- 2a ---- Electrode 2 (Po-
tential) ---- 2a ---- Electrode 3 (Potential) ---- 2a
---- Electrode 4 (Current)
                  |
    Shallow investigation (up to 5-10 meters depth)
```

- **Stage 2: Deeper Investigation (Larger Spacing)**

```scss
Copy code
                  |
    Electrode 1 (Current) ---- 4a ---- Electrode 2 (Po-
tential) ---- 4a ---- Electrode 3 (Potential) ---- 4a
---- Electrode 4 (Current)
                  |
    Deeper investigation (up to 30-50 meters depth or
more)
```

As you increase the spacing between the electrodes, the depth of the investigation increases, and you are able to probe deeper layers of the subsurface.

4. Resistivity Profile (Graphical Representation)

The resistivity data obtained from VES can be plotted to form a resistivity profile that helps in identifying different subsurface layers, such as aquifers, clays, sands, or bedrock.

Sketch 3: Resistivity Profile vs Depth

This is a simple graph of apparent resistivity against depth, showing how resistivity changes as you increase the electrode spacing.

markdown

Copy code

Apparent Resistivity (Ω·m)

```
markdown
Copy code
Apparent Resistivity (Ω·m)
 |
 |      ________________________________        ________________
 ________________________________________
 | |                                      |      |         | |
 |
 | | Layer 1 (e.g., Clay)         |      | Layer 2   | | Layer
3 (Aquifer)     |
 | |                                      |      | (e.g.,    | |
 |
 | |______________________________|      | Sand)       |
 |______________________________|
 |
 |________________________________ Depth (m)
     (0-10m)        (10-30m)     (30-50m)       (50m+)
```

- **Layer 1 (Shallow Layer)**: High resistivity, indicating a dry or less conductive layer (e.g., dry soil or rock).

- **Layer 2 (Intermediate Layer)**: Moderate resistivity, indicating possible clay or silt (a water-retaining but non-conductive layer).

- **Layer 3 (Aquifer)**: Low resistivity, indicating water-saturated zones (aquifers), where groundwater is likely present.

By interpreting these resistivity values, you can map out different geological formations and groundwater-bearing zones at varying depths.

5. VES in a Real-World Example: CrossSection of Subsurface Layers

In a typical VES survey, the results are interpreted to produce a cross-sectional view of the subsurface. This helps identify the **depth** and **extent** of different layers (e.g. aquifers, clay layers, etc.).

Sketch 4: CrossSection of Subsurface Layers Based on VES

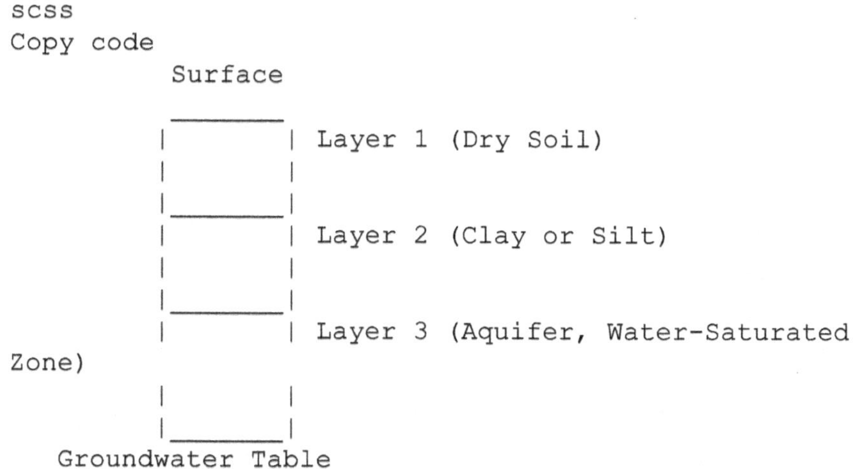

```
scss
Copy code
          Surface
         __________
        |          |  Layer 1 (Dry Soil)
        |          |
        |__________|
        |          |  Layer 2 (Clay or Silt)
        |          |
        |__________|
        |          |  Layer 3 (Aquifer, Water-Saturated
Zone)
        |          |
        |__________|
     Groundwater Table
```

- **Layer 1**: Dry soil or rock, which is non-conductive and has a higher resistivity.

- **Layer 2**: Clay, which may hold water but has higher resistivity compared to an aquifer.

- **Layer 3**: Aquifer, where resistivity drops suddenly due to low water content, making it the most desirable target for groundwater extraction.

6. Real-Time VES Survey Setup (Field View)

In a typical field setup, the resistivity equipment will be connected to the electrodes. The survey is conducted by placing the electrodes along the survey line and progressively increasing their spacing.

Sketch 5: Field Setup for VES

```
scss
Copy code
                |                        |
    [Electrode 1] ---- a ---- [Electrode 2] ---- a ----
[Electrode 3] ---- a ---- [Electrode 4]
                |                        |
```

<pre>
            (Current)   (Potential)
(Potential)   (Current)
</pre>

- **Current Injection**: Outer electrodes (1 and 4) inject current.

- **Voltage Measurement**: Inner electrodes (2 and 3) measure voltage.

- The survey is carried out by progressively increasing the electrode spacing along the survey line.

7.3.3. Summary

These sketches represent the basic concepts and field setup for **Vertical Electrical Sounding (VES)**. VES is an effective method for investigating subsurface conditions, determining the **depth of groundwater**, and identifying **aquifers**. The increasing electrode spacing allows for progressively deeper investigations, providing insights into the vertical structure of the Earth's subsurface.

7.4. Conclusion

Resistivity surveys are an excellent tool for detecting groundwater in the field, particularly in areas where drilling is either too expensive or logistically challenging. The success of the survey depends on proper planning, careful data collection, and accurate interpretation of the resistivity profiles to identify the presence of water-bearing formations.

1.In India, **Wenner and Schlumberger Arrays** are the most widely used methods for **groundwater exploration**. The **Wenner Array** is particularly favoured for **shallow investigations** due to its simplicity, while the **Schlumberger Array** is preferred for deeper exploration and high-resolution surveys. The **VES** method and **2D/3D resistivity imaging** are also gaining traction in more advanced surveys, especially in complex geological areas.

The choice of method largely depends on the specific needs of the survey, such as the **depth of investigation**, **geological conditions**, and **resolution required**.

2. Vertical Electrical Sounding (VES) is a powerful and widely used method for groundwater exploration, particularly in regions with limited access to other data sources. It provides critical insights into the depth and nature of groundwater resources, making it a valuable tool for sustainable water management in India.

◆ ◆ ◆

Chapter 8

"Borewells Unleashed: Tools and Techniques for Accessing Groundwater"

8.1. Comparison Summary Plus Exploration Chart Among Various Borewells

- Process, Depth, advantages.

- Limitations of traditional methods.

8.2. Engineering Practices in Bore well Flushing, Hydrofracturing and Blasting:

- Steps for these methods.

- Benefits and considerations.

8.3. The Case Studies in Hydrofracturing Engineering:

- A case study, Village Chandrakhuri, Block Durg, District Durg, Chhattisgarh, India

❖ ❖ ❖

After selecting the proper site resistivity methods, borehole drilling is done through any of the following methods as per the suitability of the site.

8.1. Comparison Summary plus exploration chart among various boreholes

Method	Process	Depth	Advantages	Limitations
Hand Dug Wells	Manual excavation of a hole	5–20 m (16–65 ft)	Simple, cost-effective	Shallow, prone to contamination
Rotary Drilling	Rotating drill bit with fluid circulation	Up to 3,000 m (10,000 ft)	Fast, deep, and versatile	Expensive, requires skilled operators
Percussion	Repeated dropping of a heavy bit	Up to 300 m (1,000 ft)	Effective in hard rock, medium-depth bores	Slow, water-intensive
DTH Drilling	Pneumatic hammer with compressed air	Up to 500 m (1,600 ft)	Precise, fast, and ideal for hard rock	Costly, requires air compressors and expertise
Sludging Method	Manual reciprocation of pipes with water	Up to 25 m (80 ft)	Inexpensive, good for soft soil	Shallow, ineffective in rocky terrains

Each method has specific applications, and the choice depends on depth requirements, geological conditions, and available resources.

8.2. Engineering practices for bore well Flushing

- **Flushing** a bore well is a common technique used to restore or enhance its yield by cleaning the bore well and removing blockages that might restrict water flow. Over time, bore wells can become clogged due to silt, sand, mineral deposits, or biological growth, reducing water output. Flushing helps rejuvenate the well and can sometimes improve its water-yielding capacity.

8.2.1. Steps for Flushing a Borehole

1. **Preparation**

   - **Inspection**: Use a bore well camera to inspect the well for blockages or damage. Identify the depth of the clog and its nature.

   - **Identify the water source**: Arrange an external water supply or air compressor if needed.

2. **Flushing Process**

   - **Jetting or High-Pressure Water Flushing**: A high-pressure pump or water jet is used to push water through the bore well to dislodge silt, sand, and debris.

   - **Air Compressor Flushing**: Air is injected into the well using a compressor to agitate and remove sediments from the bore well walls.

   - **Dual-Purpose Flushing**: In some cases, a combination of water and air flushing is used for more effective cleaning.

3. **Removal of Sediments**

   - As the flushing process progresses, sediment and debris are pumped out of the borewell.

   - This water is discarded away from the bore well site to avoid recontamination.

4. **Chemical Treatment (if required)**

   ○ If there are mineral deposits or biological blockages (e.g., algae), appropriate non-toxic cleaning chemicals can be introduced and flushed out later.

5. **Post-Flushing Testing**

   ○ Conduct a yield test to measure the improved water flow rate.

   ○ Collect water samples to test for quality to ensure no contamination occurred during the process.

8.2.2. Benefits of Borewell Flushing

- Restores the original water yield.

- Removes sediment, rust, and biofouling that may clog the well.

- Increases efficiency of the pump due to reduced resistance.

8.2.3. Considerations

- Flushing doesn't guarantee an increase in yield if the aquifer itself is depleted or if geological factors limit water availability.

- Professional services are recommended to prevent damage to the borewell structure.

- Over-flushing or improper techniques may damage the bore or alter its stability.

◆ ◆ ◆

Fig. 12 Hydrofracturing of Tube Wells, Village Chandrakhuri, BlockDurg (Chhattisgarh, India)

8.3. Case Study 6: Successful Hydrofracturing of a Tube Well in Chandrakhuri Village, Durg District, Chhattisgarh, India

8.3.1. Background

Water scarcity has been a persistent issue in Chandrakhuri village, located in the Durg district of Chhattisgarh, India. The region is characterised by semi-arid climatic conditions and hard rock geology, which often leads to low groundwater potential. Tube wells serve as the primary source of water for both drinking and irrigation purposes. However, many tube wells in the area fail to yield sufficient water due to aquifer limitations, leading to challenges in meeting the growing water demands.

To address these challenges, hydrofracturing—a proven technique for enhancing groundwater yield in low-yielding or dry tubewells—was implemented in Chandrakhuri village.

8.3.2. Objective

The primary objective of the project was to improve the yield of an existing tube well in Chandrakhuri village by applying hydrofracturing techniques, ensuring sustainable water availability for local residents and agricultural activities.

8.3.3. Technical Details

- **Site Location:**Chandrakhuri Village, BlockDurg, DistrictDurg, Chhattisgarh, India.

- **Geological Setting:** The area is predominantly underlain by fractured and weathered basaltic rock formations.

- **Tube Well Specifications**

  - Depth: 70 metres

  - Initial Yield: 300 litres per hour (low yield, insufficient for local needs)

8.3.4. Hydrofracturing Process

Hydrofracturing was carried out using specialised equipment to create artificial fractures in the aquifer and enhance the connectivity of existing fractures. The process involved the following steps:

1. **Assessment and Preparation**

   - Geological surveys and geophysical investigations were conducted to identify the most promising aquifer zones.

   - The tube well was cleaned to remove any debris or blockages.

2. **Injection of High-Pressure Water**

   - High-pressure water was injected into the tube well using hydrofracturing equipment.

   - The process created artificial fractures in the hard rock aquifer, improving the permeability and allowing better water flow.

3. **Post-Fracturing Development**

   - The well was further developed using air-lifting techniques to remove fine particles and ensure the effectiveness of the fractures.

8.3.5. Results

- **Post-Hydrofracturing Yield:** The yield of the tube well increased significantly to 1,500 litres per hour, a five-fold improvement.

- **Water Quality:** The water quality remained suitable for drinking and irrigation purposes, as confirmed by laboratory testing.

- **Impact:** The enhanced yield ensured sufficient water availability for approximately 50 households and nearby agricultural fields.

8.3.6. Challenges and Mitigation

- **Challenge:** Difficulty in identifying suitable zones for hydrofracturing due to the heterogeneous nature of the basaltic aquifer.

  - **Mitigation:** Advanced geophysical tools were used to pinpoint the most promising aquifer zones.

- **Challenge:** Risk of equipment damage due to hard rock formations.

- ○ **Mitigation:** Use of robust and specialisedhydrofracturing tools designed for hard rock conditions.

8.3.7. Conclusion

The successful hydrofracturing of the tube well in Chandrakhuri village demonstrates the effectiveness of this technique in enhancing groundwater yields in hard rock areas. This case study serves as a model for similar regions facing water scarcity challenges. The project not only addressed immediate water needs but also contributed to the socio-economic development of the community by supporting agricultural productivity and improving living standards.

8.3.8. Recommendations

1. Replication of hydrofracturing in other low-yielding tube wells in the region.

2. Regular monitoring and maintenance of the tube well to ensure long-term sustainability.

3. Community awareness programmes on the efficient use of water resources.

8.3.9. Acknowledgements

The successful execution of this project was made possible through the collaboration of local authorities, groundwater experts, and the community of Chandrakhuri village. Special thanks to the technical team for their expertise in hydrofracturing operations.

◆　◆　◆

8.4. Hydrofracturing of Tube Wells: Importance, Methods, and Conclusions

8.4.1. Importance of Hydrofracturing

Hydraulic fracturing, or hydro-fracking, is a method used to enhance the yield of low-performing or dry tube wells by increasing the permeability of the surrounding aquifer. This technique is particularly valuable in areas where groundwater sources are insufficient due to dense rock formations or clogged aquifer systems.

Key reasons for its importance include:

1. **Enhancing Well Productivity**: It improves water flow into the well by creating fractures or widening existing ones in impermeable or semi-permeable rock formations.

2. **Cost-Effectiveness**: Hydro fracturing can save costs by revitalising existing wells rather than drilling new ones.

3. **Addressing Scarcity**: Useful in water-scarce regions where tapping deeper or alternative aquifers is challenging.

4. **Environmental Benefits**: Reduces the need for over-drilling and limits disturbance to aquifers, contributing to sustainable groundwater management.

8.4.2. Methods of Hydrofracturing

The hydrofracturing process involves controlled injections of pressurised water into the well to create or enhance fractures in the surrounding rock. Below are the key steps:

1. **Preparation**

   - Assess the geological structure and identify the well's depth and aquifer type.

   - Clean the well to ensure unobstructed access to the desired depth.

2. **Installation of Equipment:**

   - Lower a packer (a device that seals the wellbore) into the tube well at the target fracture depth.

   - Ensure proper placement to prevent backflow and contain the pressure.

3. **Injection Process:**

   - Pump high-pressure water or a water-sand mixture into the well.

   - The pressure creates fractures in the surrounding rock or enlarges existing ones, allowing water to flow more freely into the well.

4. **Monitoring:**

   - Measure pressure and flow rate during the process to avoid excessive fracturing that could damage the aquifer.

5. **Post-Fracturing Assessment:**

   - Conduct a yield test to evaluate the effectiveness of the intervention.

   - Ensure the well›s structural integrity before resuming usage.

8.4.3. Advantages

- Enhances groundwater availability in dry or low-yield areas.

- Prolongs the lifespan of existing wells, reducing the need for new infrastructure.

- Environmentally friendly when applied judiciously.

8.4.4. Challenges

- Not all geological formations are suitable for hydrofracturing, such as unconsolidated soils.

- Requires expertise and careful pressure management to prevent over-fracturing or contamination of the aquifer.

8.4.5. Future Scope

Integrating hydrofracturing with modern hydrogeological studies and technologies can further optimise water resource management, particularly in regions prone to drought or water scarcity.

◆ ◆ ◆

8.5. Blasting of Tube Wells for Increased Yield: Importance, Methods, and Conclusions

8.5.1. Importance of Blasting

Blasting of tube wells is a traditional method used to improve the yield of wells in hard rock or dense geological formations. The technique involves creating controlled explosions to fracture surrounding rocks, increasing the permeability of the aquifer and enabling greater water flow.

Key points of importance include:

1. **Enhancing Water Yield**: Blasting can improve water flow in wells that have low productivity due to compact rock or clay layers obstructing aquifer recharge.

2. **Cost-Effective Solution**: It can restore or improve the productivity of existing wells, reducing the need for expensive new drilling operations.

3. **Use in Remote Areas**: Particularly valuable in areas lacking advanced hydrofracturing technologies, offering a simple and accessible alternative.

4. **Addressing Water Scarcity**: In water-stressed regions, this technique can provide a lifeline by increasing the yield of underperforming wells.

8.5.2. Methods of Blasting Tube Wells

1. **Site Evaluation:**

   - Conduct a geological survey to assess the surrounding rock structure and determine the suitability of blasting.

   - Ensure the aquifer and well casing can withstand the stresses from the explosion.

2. **Preparation of the Well:**

   - Remove existing equipment, debris, or obstructions from the well.

   - Determine the depth and exact location for placing the explosive charges.

3. **Placement of Explosives:**

   - Carefully lower explosives into the well to the desired depth.

   - Use a detonator and ensure the charges are securely placed.

4. **Controlled Blasting:**

   - Execute the blast under strict safety protocols, ensuring that the energy is directed towards fracturing the surrounding rock.

   - Monitor for structural integrity of the well and nearby infrastructure.

5. **Post-Blasting Procedures:**

   - Flush the well to remove debris and loose rock fragments caused by the explosion.

   - Conduct yield tests to measure the improvement in water flow.

III. Advantages

- **Increased Yield**: Significant improvement in water flow in hard rock areas.

- **Cost-Effective**: Often less expensive than advanced technologies like hydrofracturing.

- **Simple Equipment**: This does not require high-end machinery, making it suitable for rural areas.

IV. Disadvantages

- **Risk of Damage**: Uncontrolled blasting may damage the well casing or surrounding infrastructure.

- **Environmental Concerns**: Potential for aquifer contamination or unintended fractures.

- **Limited Suitability**: Ineffective in soft or unconsolidated formations.

V. Recommendations

- Use blasting only after a professional assessment of the geological conditions.

- Employ skilled personnel to execute controlled explosions.

- Combine blasting with modern groundwater management techniques for sustainable water extraction.

Blasting remains a valuable tool for enhancing well yields, particularly in resource-limited or remote areas, provided it is applied judiciously and with appropriate safeguards.

8.6. Conclusions

Flushing of bore wells is used to remove sediment, ensure water quality, and maintain flow. Use proper techniques, prioritise safety, dispose of waste responsibly, and test water quality post-flushing for optimal performance.

Hydrofracturing of tubewells has proven to be an effective technique for improving groundwater extraction in challenging geological settings. However, its success depends on proper site evaluation, skilled execution, and post-fracturing monitoring.

Blasting is a practical and historically proven method to enhance tube well productivity in challenging geological conditions. However, it requires careful planning and expertise to achieve the desired results while minimising risks.

◆ ◆ ◆

Chapter 9

Innovative Recharge: The V-Wire Solution for Sustainable Groundwater Use

9.1. V-Wire Technology for Recharge at Village Main Drain Outlet: Principles, Methods, Advantages, and Conclusions

- Principles, technical methods, Aquifers, Advantages.

- Conclusion and Key takeaways'.

9.2. Schematic diagram for V-wire technology:

- Pictures of methods.

- The role of divining.

9.3. The case studies:

- Bemetara village Umaria.

- GajaraWatershed, DistrictDurg.

❖ ❖ ❖

9.1. V-Wire Technology for Recharge at Village Main Drain Outlet: Principles, Methods, Advantages, and Conclusion

9.1.1. Principles of V-Wire Technology

V-wire technology is a system used to enhance groundwater recharge while filtering impurities from water. It is particularly effective at village drain outlets, where wastewater can be treated and directed back into the aquifer. The method uses a *V-shaped wire screen filter* to prevent clogging while allowing water infiltration.

The primary principles include:

1. **Filtration and Percolation**: The V-wire design ensures efficient filtration of suspended solids, allowing only cleaner water to pass through for recharge.

2. **Sustainability**: Converts village drain outflows, typically a source of contamination, into a recharge mechanism for groundwater.

3. **Controlled Infiltration**: Promotes gradual and controlled infiltration, protecting the aquifer from overloading or contamination.

9.1.2. Technical Methods

1. **Site Selection**:

   - Identify a suitable location at the main drain outlet near the village, ensuring proximity to an aquifer.

   - Assess soil permeability and hydrogeological conditions for recharge potential.

2. **Construction of Recharge Structure**:

   - Excavate a recharge pit or trench at the outlet site.

- Line the pit with a *V-wire screen* (stainless steel wires welded in a V-shape) to filter solid particles.

- Install layers of graded gravel and sand below the screen to provide additional filtration and prevent clogging.

3. **Pre-Treatment of Water**:

- Divert the drain water into a sedimentation tank to remove heavy debris and silt.

- Optionally, include treatment steps like bioremediation or slow sand filtration for enhanced water quality.

4. **Recharge Mechanism**:

- The filtered water flows through the V-wire screen and infiltrates into the ground, reaching the aquifer.

- Use monitoring wells to measure recharge efficiency and ensure aquifer health.

5. **Maintenance**:

- Regularly clean the V-wire screen and remove accumulated debris from the sedimentation tank.

9.1.3. Advantages of V-Wire Technology

1. **Efficient Filtration**:

- The V-shaped design prevents clogging, allowing for uninterrupted water infiltration.

2. **Enhanced Groundwater Recharge**:

- Converts unused or wastewater into a valuable resource by replenishing aquifers.

3. **Cost-Effective**:

- Requires minimal infrastructure compared to advanced recharging technologies.

4. **Improves Water Quality**:

   ◦ Reduces contaminants through multiple filtration layers, protecting aquifer quality.

5. **Sustainable Drainage Management**:

   ◦ Provides a dual solution by managing wastewater and combating groundwater depletion.

6. **Community Impact**:

   ◦ Promotes water security in villages, aiding agriculture and domestic water use.

9.1.4. Role of Water Divining in Injection Wells Location

Water divining, also known as dowsing, is a traditional method used to locate underground water sources. While it is often considered a pseudoscientific practice, some individuals use it to guide decisions in water resource management. In the context of an **injection well for a V-wire recharging system**, the potential role of water divining could involve:

1. **Identifying Aquifer Locations**:

   Practitioners might use divining techniques to locate the ideal spot for an injection well. The aim is to identify areas with the highest likelihood of a permeable aquifer where the recharge process can be most effective.

2. **Assessing Subsurface Features**:

   Water divining might be employed to predict subsurface features, such as fractures or porous soil layers, which are critical for the functioning of a V-wire recharging system.

3. **Complementing Geological Surveys**:

   Although not a scientifically reliable method, water divining might be used alongside modern techniques, such as geophysical surveys, to cross-check or affirm findings before drilling.

9.1.5. Scientific Perspective

The effectiveness of water divining is not supported by empirical evidence, and its success is often attributed to chance or the dowser's familiarity with local hydrogeological conditions. Modern approaches like **geophysical surveys (e.g., resistivity or seismic methods)** and **hydrogeological modelling** are more accurate and reliable for designing and siting injection wells.

9.1.6. V-Wire Recharging System Context

In a **V-wire recharging system**, an injection well is designed to transfer water (e.g., rainwater runoff) back into an aquifer. Its performance depends on:

- **Proper site selection**: Based on soil permeability and aquifer characteristics.

- **Design efficiency**: Ensuring the V-wire screens resist clogging and facilitate efficient water flow.

Instead of relying on water divining, using scientific methods such as borehole logging, pump testing, and groundwater modelling is recommended to ensure the optimal performance of the recharging system.

*Figs. 13, 14, 15, 16, 17 & 18 All for V-Wire Recharging work,
Block-Bemetara, Chhattisgarh, India*

9.2. The CASE Studies

9.2.1. Case Study 7: V-Wire Recharge System at Main Drain Outlet, Villages Umaria&Piparhatta

1. Introduction

- **Location**: village Umaria, village Piparhatta, Bemetara district, Chhattisgarh.

- **Need for Recharge System**: The main drain outlet likely carries a significant volume of water, often lost as surface runoff, which can be harnessed for groundwater recharge.

- **Objective**: To design and implement a V-wire recharge system to enhance groundwater levels and mitigate water scarcity.

2. Site Characteristics

- **Drainage Profile**:
  - Drain source: Village runoff, agricultural return flows, and possibly domestic wastewater.
  - Flow pattern: Seasonal or perennial.
  - Water quality: Assessment of pollutants, sediment load, and suitability for recharge.

- **Hydrogeological Context**:
  - **Soil Type:** Sandy, clayey or loamy, determining the infiltration capacity.
  - **Aquifer Type:** Presence of unconfined or confined aquifers, depth, and permeability.
  - **Water Table:** Baseline groundwater levels to measure recharge effectiveness.

- **Land Use:**
  - Surrounding agricultural activities or infrastructure.
  - Available area for installing the recharge system.

3. Design of V-Wire Recharge System

- **System Components:**
  - **V-Wire Screen**: Designed to filter debris and prevent clogging, ensuring smooth percolation of water into the aquifer.
  - **Injection Well:** Drilled to a depth that connects to the target aquifer layer.
  - **Pre-Treatment Chamber:** To settle sediments and remove contaminants before water enters the system.

- **Capacity Estimation:**
  - Volume of water from the drain during peak flow.
  - Recharge rate based on aquifer permeability and screen efficiency.

- **Quality Management:**
  - Inclusion of a biofiltration layer or sedimentation tank for initial water treatment to prevent aquifer contamination.

4. Implementation Steps

1. **Baseline Surveys:**
   - Topographical survey to understand the drainage gradient.
   - Geophysical surveys to locate suitable aquifer zones.

2. **System Construction:**
   - Drilling and installation of the injection well.
   - Placement of V-wire screens and supporting components.

3. **Monitoring Setup:**

   - Install piezometers to monitor groundwater level changes.

   - Regular water quality checks.

5. Outcomes and Benefits

- **ShortTerm:**

  - Reduction in surface water runoff.

  - Improved water quality at the drain outlet.

- **Long-Term:**

  - Increased groundwater recharge benefits local wells and agriculture.

  - Mitigation of water scarcity during dry seasons.

- **Community Impact:**

  - Potential improvement in agricultural productivity.

  - Reduced dependency on external water sources.

6. Challenges

- **Water Quality**: High levels of contaminants in drain water may require additional treatment.

- **Community Awareness**: Ensuring local participation and understanding of system benefits.

- **Maintenance**: Regular cleaning of V-wire screens and pre-treatment chambers to maintain efficiency.

7. Recommendations

- Integration with a **rainwater harvesting plan** to increase recharge potential.

- Collaboration with local governing bodies and community for long-term sustainability.

- Periodic monitoring and evaluation to refine the system design.

9.2.2. Schematic Diagram of Innovative V-Wire Injection Well

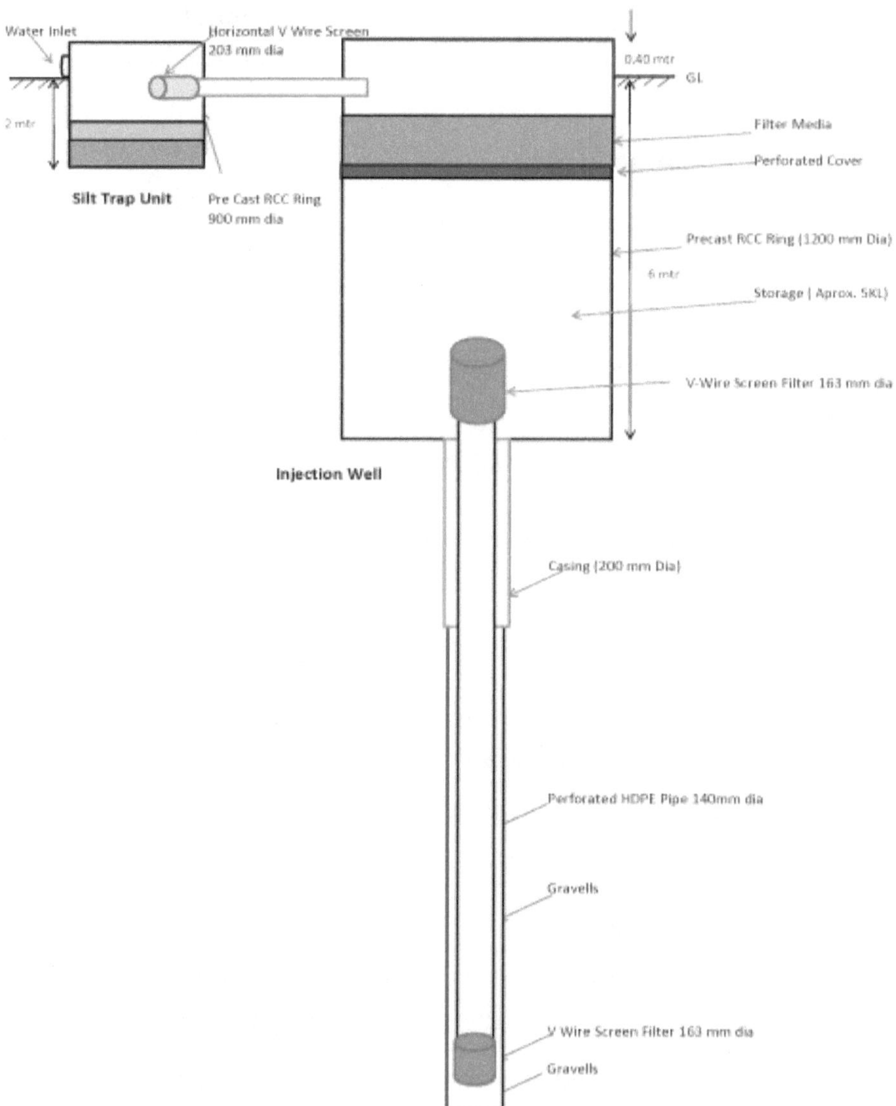

Fig. 19 V-Wire Technology, Schematic Diagram.

9.2.3. Case Study 8: Groundwater Recharge Success Stories in Gajranala Watershed, Durg District, Chhattisgarh

1. Introduction

- **Location**: The Gajra Watershed lies in the Durg District of Chhattisgarh, a region with mixed agricultural and industrial activities.

- **Problem Statement**:

  - Groundwater depletion due to over-extraction for irrigation and domestic use.

  - Reduced natural recharge due to changes in land use and erratic rainfall.

- **Objective**: To evaluate the potential for groundwater recharge through watershed management practices.

2. Study Area Description

- **Geography**:

  - Total area of the Gajra Watershed is 250 sq km. The year of construction was 2003, with a total number of 46 villages in the Patan block of Durg district.

  - Topography: Dominantly undulating terrain with drainage patterns flowing towards seasonal rivers.

- **Climatic Conditions**:

  - Average Annual rainfall: 1147 mm, monsoon-dominated, with variability affecting recharge potential.

  - Temperature and evaporation rates impact water retention.

- **Land Use:**
  - Predominantly agricultural, with some forested and industrial areas.
  - Types of crops and irrigation methods used.
- **Hydrology:**
  - Seasonal streams and drains contribute to the watershed.
  - Natural recharge points like ponds, depressions, and riverbanks.

3. Hydrogeological Context

- **Aquifer Characteristics:**
  - Type: Hard rock aquifers, Charmuria, Gunderdehi and Chandi Formations, limestones and shales, typical of Chhattisgarh, with fractures and weathered zones.
  - Depth to water table: Pre-monsoon and post-monsoon variations.
  - Storage capacity: Limited due to rocky terrain but potential in weathered zones.
- **Soil and Permeability:**
  - Dominant soil type: Black cotton soil with low permeability in some areas.
  - Recharge potential: Enhanced in sandy loam areas and fractured zones.
- **Groundwater Extraction:**
  - Major uses: Agriculture, domestic, and industrial.
  - Over-extraction hotspots identified using historical water level data.

4. Groundwater Recharge Interventions

- **Existing Practices:**
  - Traditional structures like 100 ponds and tanks.
  - Check dams and bunds constructed by local authorities.
- **Proposed/Existing Interventions:**

1. **Rainwater Harvesting:**
   - Rooftop collection in villages and small towns.
   - Surface runoff harvesting using contour trenches and bunds.

2. **Percolation Tanks:**
   - 25 nos. Creating tanks near fractured zones for maximum infiltration.

3. **Check Dams Stop Dams and NalaBunding:**
   - Installing 35 small & medium check dams, 28 masonry stop dams on seasonal streams to slow down runoff and allow percolation, and 22 subsurface dykes.

4. **V-Wire Injection Wells:**
   - Using boreholes fitted with filters proposed but not used here to directly recharge deeper aquifers.

5. **Rejuvenation of Traditional Systems:**
   - Desilting and repairing old tanks, ponds, and 40 nala bunds to restore capacity.

5. Implementation Strategy

1. **Survey and Planning:**
   - Mapping of the watershed using GIS tools to identify recharge zones.

- Socio-economic survey to understand water use and stakeholder priorities.

2. **Construction and Installation:**

- Collaboration with local agencies and communities for implementation.

- Use of local materials and labour for cost efficiency.

3. **Monitoring and Maintenance:**

- Regular monitoring of water levels using piezometers.

- Maintenance of recharge structures, such as desilting tanks and repairing cracks in check dams.

6. Outcomes

- **Hydrological Benefits:**

  - Increased groundwater levels, especially during post-monsoon.

  - Reduction in runoff and soil erosion.

- **Agricultural Benefits:**

  - Improved availability of water for irrigation, leading to higher crop yields.

  - Reduced dependence on boreholes, lowering energy costs for farmers.

  - Command Area enhanced from 7 sq km to 20 sq km, power pump numbers increased from 690 in 2003 to 5500 in 2019 & pumping hours now range from 6-10 hrs compared to 3-5 hrs in the year 2003.

- **Environmental Benefits:**

  - Rejuvenation of local ecosystems, including wetlands and stream flows.

- Improved resilience to droughts and water scarcity.
- Hand pumps 436 increase to 2079, 17-31m groundwater level upgraded to 4-12 m.

7. Challenges

- **Community Participation:**
  - Initial resistance due to lack of awareness about recharge benefits.

- **Technical Limitations:**
  - Hard rock terrain posing challenges for infiltration.

- **Financial Constraints:**
  - High initial investment for constructing large-scale recharge structures.

8. Recommendations

- **Capacity Building:**
  - Training programmes for farmers and local leaders on sustainable water management.

- **Integration with Government Programmes:**
  - Leveraging schemes like **Mahatma Gandhi National Rural Employment Guarantee Act (MGNREGA)** and watershed development projects.

- **Long-Term Monitoring:**
  - Setting up a database for groundwater levels and quality to measure impact.
  - Uses a number of V -wire injection wells at certain specified locations may enhance the rate of groundwater recharges.

9.3. Key Takeaways

V-wire technology is

- Simple yet robust design suitable for rural infrastructure.

- Improves groundwater levels while addressing wastewater management challenges.

- Requires regular maintenance but offers long-term benefits for water security.

This technology exemplifies a sustainable approach to water resource management, combining innovation with simplicity to empower communities in water-stressed regions.

◆ ◆ ◆

9.4. Conclusions

V-wire technology is a practical and effective solution for addressing groundwater depletion while managing wastewater from village drain outlets. Its integration of filtration, recharge, and sustainability makes it an ideal choice for rural settings. A V-wire recharge system at the main drain outlet of Umaria&Piparhatta can serve as a model for sustainable water management in rural Chhattisgarh. By harnessing drain water, the system can address groundwater depletion issues while supporting agriculture and domestic water needs. Proper planning, community involvement, and ongoing monitoring are essential for success.

The Gajra Watershed in Durg District holds significant potential for groundwater recharge. By implementing targeted interventions such as percolation tanks, check dams and injection wells, the region can combat groundwater depletion and support sustainable agriculture and livelihoods. A participatory approach involving the local community and government agencies is critical for the long-term success of recharge efforts.

◆ ◆ ◆

Chapter 10

The Ethical and Practical Considerations of Divining in Modern Water Solutions

10.1. Scientific Scepticism vs. Traditional Knowledge:

- The scientific community's views on divining and why it's often considered unproven or pseudoscience.

- How cultural and contextual factors influence perceptions of divining.

10.2. Building Trust with Local Communities:

- How integrating divining into water projects can help build trust between engineers, NGOs, and local communities.

- Overcoming cultural resistance to modern engineering practices and vice versa.

◆ ◆ ◆

10.1. Scientific skepticism vs. Traditional Knowledge

Divining, or dowsing, is the practice of locating underground water sources using tools like dowsing rods or pendulums, or through intuitive means. While deeply rooted in tradition, particularly in rural and indigenous communities, divining is often met with scepticism in the scientific community. Modern water management strategies, particularly those relying on engineering and geophysical methods, typically do not integrate divining due to the lack of scientific evidence supporting its efficacy. However, the integration of divining with modern water solutions has sparked debate about the ethical and practical considerations of balancing **scientific scepticism** with **traditional knowledge**.

This section explores the scientific community's views on divining, why it's often considered unproven or pseudoscience, and how cultural and contextual factors influence perceptions of divining in different regions.

1. Scientific Scepticism vs. Traditional Knowledge

10.1.1. Scientific Communities view

10.1.1.1. Scientific Scepticism Towards Divining

The scientific community generally dismisses **divining** as pseudoscience. There are several reasons why divining has not gained acceptance within mainstream scientific practice:

1. **Lack of Empirical Evidence**: One of the primary reasons dowsing is seen sceptically is the absence of **consistent, reproducible evidence** that proves it works. Scientific methods emphasise **control, measurement, and repeatability**, and so far, there has been no study showing that dowsing can predict groundwater with greater accuracy than chance or random guessing.

   o **Controlled Studies**: Numerous controlled experiments have been conducted in an attempt to test the accuracy of divining, but the results have generally shown that diviners perform no better than random chance. For example, studies in the U.S., Europe, and other parts of the world have tested diviners against controls, and findings have consistently shown no statistically significant difference in success rates.

2. **Lack of a Mechanism**: From a scientific standpoint, there is no clear or plausible mechanism to explain how divining might work. Modern science relies on measurable physical phenomena, such as **electromagnetic fields**, **geophysical properties**, and **hydrological conditions**, to predict where water may be located. Divining, on the other hand, is based on intuition or the use of tools like rods or pendulums, none of which can be linked to scientifically understood forces that could influence groundwater.

3. **Psychological and Cognitive Biases**: Many scientific critiques of divining cite **cognitive biases**, such as the **confirmation bias** or the **ideomotor effect** (the unconscious movement of the dowsing rod or pendulum), as possible explanations for its perceived success. According to this view, diviners may unconsciously move the rods in response to subtle cues from the environment or their own expectations, which can give the illusion of detecting water.

4. **Placebo Effect**: In some cases, dowsing might work due to the **placebo effect** or the belief that dowsers are skilled in finding water. People may be more likely to **invest in water-sourcing efforts** when they believe dowsing is successful, even if the success is based on luck or coincidence.

10.1.1.2. Why Some Communities Value Divining

Despite the scepticism in scientific circles, divining remains an important and often trusted tool for locating water in many regions, particularly in **rural** and **developing** areas. This trust in divining is shaped by several cultural and contextual factors:

1. **Cultural Significance**: In many communities, divining has **deep cultural roots** and is seen as an ancient skill passed down through generations. For these communities, divining is not just a method of finding water but is part of their **cultural identity**. Diviners are often respected members of society, and their methods are ingrained in local traditions, making it difficult for scientific scepticism to penetrate these practices.

   ○ **Local Belief Systems**: In rural or indigenous cultures, people often have a **holistic view of nature** and believe that humans have an innate connection with the natural world. Divining is seen as tapping into that connection—an understanding that science, with its focus on data and objectivity, may not always account for.

   ○ **Intergenerational Knowledge**: Knowledge of local water sources, passed down from ancestors or elder community members, plays a key role in how divining is viewed. The idea that water divining is part of this tradition lends it credibility, even in the absence of scientific validation.

2. **Contextual Effectiveness**: In areas with **limited access to modern technology**, such as remote rural locations, divining can provide practical solutions when more advanced tools are unavailable. In these areas, diviners are often seen as experts who can help locate water in regions where the geological composition is complex, or where scientific surveys are difficult to carry out due to economic constraints or lack of infrastructure.

   ○ **Economic Considerations: In places where geophysical equipment and specialised expertise are too costly, diviners may offer a more affordable alternative for initial exploration. Divining can be an accessible first step, helping communities or engineers to identify**

potential sites before committing to expensive drilling or more comprehensive geophysical assessments.

3. **Practical Success in Specific Conditions**: In certain contexts, divining has been reported to have **practical success** in locating water. Although this success is often anecdotal, it is **culturally reinforced** over time. In some regions, diviners have pointed to aquifers or water pockets that engineers or hydrologists had not identified, particularly in **difficult or uneven terrain**.

4. **Community Trust**: Diviners are often trusted local figures who have built their reputations over years of practice. In rural areas where trust in outside experts or engineers may be low, diviners can act as **mediators** between the community and external water management efforts. Their involvement in water planning projects can encourage **community buy-in** and enhance local cooperation.

10.1.2. Cultural and Contextual Factors Influencing Perceptions of Divining

10.1.2.1. Influence of Local Experience and Environmental Context

The perception of divining varies depending on environmental conditions, historical experiences with water scarcity, and cultural perspectives on nature and science.

1. **Historical Water Scarcity**: In regions that experience periodic droughts or have limited access to clean water, divining may be seen as a valuable tool in ensuring water security. In areas where engineering methods have failed or are seen as ineffective, divining may gain credibility as an alternative, especially when it seems to "work" in locating water.

2. **Rural vs. Urban Divide**: In **urbanised** or **highly developed** regions, scientific methods for water management dominate, and divining is more likely to be dismissed. However, in **rural** or **isolated areas**, where access to modern tools and data is limited, divining may be more accepted as part of the local problem-solving toolkit.

3. **Religious and Spiritual Dimensions**: In some cultures, divining is closely tied to **spiritual beliefs**. The practice may be seen as a form of **sacred knowledge** or an interaction with the divine, making it more than just a technique for locating water. This connection can further influence how people perceive its effectiveness, as it may not be seen as purely practical but also metaphysical.

10.1.2.2. *Modernisation and Changing Attitudes*

As **education levels rise** and **access to technology** increases in rural regions, perceptions of divining may shift. Younger generations or those who have been exposed to scientific methodologies may be more sceptical of divining, while older generations or rural residents may continue to trust it based on longstanding cultural practices. In many places, this creates a **generational divide** in perceptions, with older individuals favouring traditional practices and younger ones leaning towards scientific approaches.

- **Hybrid Approaches**: In some cases, younger generations might be more open to **combining scientific methods** with traditional knowledge. For example, while they may acknowledge the scientific methods of geological surveys, they might still see value in consulting diviners as part of a comprehensive approach to water management.

10.1.3. Ethical Considerations in Integrating Divining into Modern Water Solutions

When considering the use of divining in modern water projects, there are several **ethical** issues that must be carefully balanced:

10.1.3.1. *Respecting Cultural Practices*

It's crucial for modern engineers and water professionals to **respect local cultural practices** and acknowledge the role divining plays in communities. While divining may lack scientific validation, its use is often deeply embedded in the culture and worldview of the people who practice it.

- **Cultural Sensitivity**: Engineers working in rural or developing areas should engage with diviners in a **culturally sensitive manner**, incorporating local traditions while still relying on scientific methods for verification and long-term water sustainability. This approach fosters **trust** and ensures that community members feel involved and respected throughout the process.

10.1.3.2. *Integrating Divining with Scientific Methods*

Ethically, it is important not to exploit the practice of divining for profit or to mislead communities into thinking it can replace scientifically proven methods. Divining should be viewed as **complementary** to modern engineering techniques, not as a substitute. The integration should be transparent and based on the **understanding that divining alone cannot guarantee the sustainability** or quality of a water source.

10.2. Building Trust with Local Communities: The Role of Divining in Water Projects

In water resource management projects, particularly in rural or developing regions, establishing **trust** between engineers, NGOs, and local communities is crucial for the success of the initiative. Communities often have longstanding traditions and practices related to water sourcing, and these must be acknowledged and respected in order to ensure cooperation and successful implementation. **Integrating divining into water projects** can serve as a bridge between traditional knowledge and modern engineering practices, helping to build this trust. However, doing so also requires addressing cultural resistance on both sides: the scepticism of modern methods in local communities, and the reluctance of engineers or NGOs to incorporate traditional practices seen as unscientific.

This section explores how **integrating divining** into water resource projects can foster trust and **overcome cultural resistance**, leading to more inclusive, effective, and sustainable solutions.

10.2.1. How Integrating Divining into Water Projects Can Build Trust

10.2.1.1. Recognising the Role of Local Knowledge

Divining is often seen as a part of **local heritage**—a knowledge system passed down through generations. When engineers and NGOs acknowledge and incorporate this practice, they demonstrate **respect for the local culture** and understanding of the community's values and practices. This helps foster trust and cooperation, as it shows that the project's goal is not just to implement an external solution, but to **work with** and **learn from** the community.

- **Community Engagement**: In rural or indigenous communities, diviners are often viewed as **trusted figures**. By involving diviners in the planning process, engineers and

NGOs send a clear message that they value the community's expertise, which can encourage local participation and cooperation. Diviners may help identify potential water sources that modern technology has overlooked, and their involvement can make community members feel more invested in the project.

- **Inclusive Decision-Making**: When diviners are invited to contribute alongside engineers and hydrologists, it becomes a **collaborative approach**, not an imposition of external ideas. This shared responsibility and respect for local knowledge can significantly improve **community buy-in** and ownership of the project, leading to better outcomes in the long run.

10.2.1.2. Building Trust Through Shared Goals

Both diviners and engineers share a common goal: to provide **clean, reliable water** to communities. The shared mission of addressing water scarcity creates an opportunity to **unite different knowledge systems** for the common good.

- **Common Purpose**: By framing the project as a partnership between local diviners and modern engineers, all parties can focus on the shared objective of ensuring a sustainable and safe water supply. This **aligns their interests**, making it easier for engineers and diviners to work together towards a common goal and creating a sense of mutual respect.

- **Community Ownership**: When communities see that their **traditional knowledge is respected and utilised**, they are more likely to support the project, view it as legitimate, and **take ownership** of its success. This trust in the process can translate into a greater willingness to invest time and resources in maintaining water systems or taking responsibility for the sustainable use of water sources.

10.2.1.3. Empowering Local Diviners and Building Local Capacity

Incorporating divining into water planning can also be a way of **empowering local communities**. Diviners are often skilled professionals in their own right, and involving them in technical projects gives them **formal recognition** for their expertise, which can enhance their standing within the community and ensure their knowledge is preserved.

- **Skill Development**: Diviners may also gain technical skills by working alongside engineers, which can help them **expand their knowledge** of modern water management techniques. This fosters a culture of **mutual learning** and growth, improving the overall capacity of local communities to manage their water resources effectively.

- **Sustainability**: When local knowledge is integrated into the project, the solution becomes more **sustainable** in the long term. Diviners can provide ongoing monitoring or advice on water availability and location, ensuring that the community is not reliant on outside experts for future water needs.

10.2.2. Overcoming Cultural Resistance to Modern Engineering Practices

10.2.2.1. Acknowledging and Respecting Cultural Beliefs

Cultural resistance to modern engineering solutions often stems from a lack of understanding, fear of change, or a perception that modern practices are **imposed by outsiders** who do not understand the local context. Engineers and NGOs must approach these concerns with cultural sensitivity, taking care not to dismiss traditional knowledge outright.

- **Open Dialogue**: A **dialogue-based approach** where engineers and NGOs explain the benefits of modern water management systems while also listening to the concerns of the community is essential. By recognising the **validity of local knowledge** and integrating it into their work, they can create an environment of mutual respect and shared decision-making.

- **Inclusive Planning**: Communities are more likely to accept engineering solutions when they feel that the project has been **designed with their input**, not just for their benefit. By including diviners in the planning stages, engineers can show that they are committed to working collaboratively, rather than imposing foreign technologies or methods.

10.2.2.2. *Addressing Scepticism About Modern Engineering*

In rural or marginalised communities, there may be **scepticism towards modern engineers** or external experts, especially if past projects have failed or if the community feels that their needs have been ignored. Divining can play a critical role in addressing this resistance, as it helps **integrate familiar practices** with new technologies.

- **Trust through Familiarity**: When diviners are involved in modern water projects, it helps to bridge the gap between **traditional wisdom** and **modern technology**. Communities are more likely to trust engineers and NGOs if they see that these outside experts are **working with their own people**. The presence of trusted local diviners can help **validate the project**, making it seem less foreign and more in tune with local realities.

- **Piloting with Diviners**: One way to overcome scepticism is to begin with a **pilot project** where divining is used in conjunction with engineering methods. By showing that divining can help identify water sources that modern methods might have missed, engineers can **demonstrate the value** of combining both approaches, rather than positioning one as superior to the other.

10.2.2.3. Transparency and Education

Part of overcoming cultural resistance involves **educating the community** about the benefits of modern water management practices and how these practices align with their traditional knowledge. Engineers and NGOs should not simply impose new technologies but should explain **why modern methods are effective** and how they can improve water security over the long term.

- **Public Workshops**: Organising **community meetings** or workshops to educate people about modern water technologies can help demystify engineering processes. In these forums, diviners can share their experiences, and engineers can explain how modern tools, such as geophysical surveys and water testing, work in tandem with divining to create more reliable outcomes.

- **Mutual Understanding: Building mutual respect** through education helps communities understand that modern engineers are not replacing traditional practices, but rather **enhancing them** with tools that make water sourcing more predictable, sustainable, and scientifically sound.

10.2.3. Overcoming Resistance to Divining from Engineers and NGOs

Just as local communities might be resistant to new engineering practices, there may also be **scepticism within the engineering community** regarding the effectiveness and scientific legitimacy of divining. This resistance can be addressed by highlighting the **practical benefits** of incorporating local knowledge, as well as the ethical obligation to engage with communities in a culturally sensitive manner.

10.2.3.1. Encouraging Collaboration and Co-Creation

Engineers and NGOs should **encourage collaboration** with diviners, not as a mere concession to local beliefs, but as a way to **create more holistic, culturally relevant solutions**. Engineers can learn from diviners' intuitive understanding of local geography and hydrology, while diviners can benefit from technical training and data provided by engineers.

- **Training Programmes**: Engineers and diviners can engage in **mutual training** to ensure that both groups understand each other's approaches. Engineers can learn to appreciate the value of divining as part of a larger, integrated solution, while diviners can gain insights into the scientific aspects of water sourcing.

10.2.3.2. Scientific Validation of Divining

One way to bridge the gap between scepticism and practical application is to **scientifically validate the role of divining** in certain contexts. While divining may not have a scientifically established mechanism, cases where it successfully locates water sources can be used as evidence that traditional knowledge can play a valuable role in water management.

- **Pilot Studies**: Engineers can collaborate with diviners on pilot studies where divining results are tested against modern geophysical methods. If divining proves useful in certain contexts, it could help to **normalise its use** in future projects, integrating it as a complementary tool rather than dismissing it outright.

10.3. Conclusion

The ethical and practical considerations of integrating divining into modern water solutions revolve around **respecting cultural knowledge** while ensuring that scientifically sound practices are upheld. While the scientific community remains sceptical about divining due to the lack of empirical evidence and a clear mechanism, local traditions and cultural practices continue to view divining as a valuable tool for water sourcing, especially in remote or underserved.

Integrating divining into modern water projects can serve as a **powerful tool for building trust** between engineers, NGOs, and local communities. By recognising and respecting the importance of local knowledge and culture, engineers can facilitate **greater community involvement**, reduce resistance to modern solutions, and ensure more **sustainable and inclusive water management**. Ultimately, the success of these projects depends on the ability to **integrate scientific methods with traditional practices**, creating solutions that are both effective and culturally relevant.

◆ ◆ ◆

Chapter 11

Tools and Techniques—A Practical Guide for Engineers and Water Professionals

11.1. How to Collaborate with Diviners:

- Guidelines for working with diviners to locate water sources.

- How diviners can complement hydrogeological surveys and remote sensing data.

11.2. Integrating Divining into Engineering Models:

- Practical approaches to using divining results alongside geological data in borewell planning.

- Using divining as a preliminary tool before committing to more expensive or invasive drilling methods.

11.3. Case Study: A Step-by-Step Example of Integration:

- Walkthrough of a project that used divining techniques as a primary or secondary tool in locating water.

◆ ◆ ◆

11.1. How to Collaborate with Diviners: Guidelines for Working with Diviners to Locate Water Sources

Working with diviners in water resource planning requires a **collaborative approach** that respects the role of traditional knowledge while integrating scientific methods to ensure accuracy, efficiency, and long-term sustainability. This partnership can offer valuable insights, especially in regions where modern technologies may face limitations due to terrain, resource constraints, or community resistance to external methods.

The following guidelines outline how engineers and water professionals can collaborate effectively with diviners to locate water sources and **enhance the success** of water projects.

11.1.1. Establishing a Respectful Relationship

11.1.1.1. Build Trust and Mutual Respect

The foundation of any collaboration with diviners is trust. Diviners are often seen as local experts with deep cultural and environmental knowledge. Engineers and water professionals should approach divining with **openness** and a willingness to learn from the experience of the diviners. Building a relationship based on **mutual respect** will pave the way for effective teamwork.

- **Acknowledge Expertise**: Recognise diviners as knowledgeable professionals within the community. This can be done by engaging with them as **partners** rather than just consultants.

- **Community Engagement**: Start by engaging with the **community as a whole**, explaining the role of diviners in the project and clarifying how their involvement will benefit the community. Publicly acknowledging the value of divining can increase local acceptance of the project.

11.1.1.2. Facilitate Open Communication

Ensure clear and transparent communication from the outset. Engineers should explain the technical aspects of their approach (e.g., hydrogeological methods, drilling techniques, etc.), while diviners can share their methods, practices, and previous successes in locating water sources.

- **Create an Inclusive Team**: The project team should include both **technical experts** (engineers, hydrogeologists) and **local diviners**. This fosters a **team-oriented mindset**, where every member's skills are valued.

- **Define Roles and Expectations**: Clearly outline the specific roles of diviners within the project. Whether their role is to **guide initial site selection** or to assist during drilling, setting expectations will help avoid misunderstandings and conflicts.

11.1.1.3. Be Open to Flexibility

Both engineers and diviners need to remain **flexible** and adaptable throughout the collaboration. Water sourcing can be unpredictable, and there might be moments when **traditional practices** or **scientific techniques** don't yield immediate results. Successful collaboration requires openness to **trial and error** and a willingness to explore new methods.

- **Complementary Process**: Understand that divining and engineering are complementary approaches. Diviners may suggest potential water sites, but engineers will need to verify these suggestions using **hydrological methods**.

11.1.2. Guidelines for Effective Collaboration with Diviners

11.1.2.1. Step-by-Step Integration of Divining in the Water Sourcing Process

1. **Initial Site Identification (Diviner›s Role):**

   - **Diviners can perform initial site selection** by using their tools (such as dowsing rods, pendulums, or their own intuition) to **identify potential water sources**. In areas where the terrain is challenging or data is sparse, divining can guide the early stages of water exploration. Diviners' knowledge of local groundwater flows and ancient wells can highlight areas with higher potential for successful borewell drilling.

   - **Community Engagement**: Diviners may already have an intuitive sense of water patterns, informed by **local knowledge** (e.g., past well locations, seasonal variations). This insight can help direct engineers to promising locations before committing significant resources.

2. **Hydrogeological Survey (Engineer›s Role):**

   - After divining suggests a site, engineers should **conduct hydrogeological surveys** to validate the diviner's findings. This includes studying soil composition, **geophysical resistivity** (using tools like ground-penetrating radar), and **drilling tests** to confirm the existence of water and its quality.

   - **Surface and Subsurface Investigation**: Engineers can also carry out **remote sensing** (e.g., satellite imagery) and **ground truthing** to supplement divining and identify the most viable drilling locations.

3. **Drilling and Water Testing (Joint Role):**

   - Once the site is identified and verified, engineers should carry out **drilling operations**. During this stage, diviners can continue to assist by **monitoring the drilling process**, ensuring the water source is consistent with their earlier predictions.

   - **Water Quality and Flow Assessment**: Engineers can then test the water for **quality, depth, flow rate**, and other parameters such as **salinity** and **pH** to ensure it meets the community's needs.

4. **Post-Implementation Monitoring:**

   - After the borewell is operational, diviners can assist with **ongoing monitoring** and assessment, providing **feedback on the sustainability** of the water source and guiding **maintenance efforts**. In areas where groundwater levels fluctuate or are influenced by seasonal changes, diviners can help **identify potential risks** or shifts in water availability over time.

11.1.2.2. Working with Diviners: Practical Tips

1. **Respect and Acknowledge Local Practices:**

   - **Provide space for diviners to perform their work** without interference. Avoid dismissing their methods outright, as this can lead to friction. Instead, incorporate their suggestions alongside other scientific techniques to create a **holistic approach**.

   - **Learn from the Local Context**: Diviners are often deeply attuned to **local environmental patterns**, such as rainfall, plant life, or animal behaviour, that may be linked to water sources. Encourage diviners to share this **contextual knowledge** with engineers to enrich the project.

2. **Ensure Collaboration at Every Stage:**

   ○ **Early Involvement:** Bring diviners in at the earliest stages of project planning, not as an afterthought. This ensures that they have time to provide input and be part of the decision-making process.

   ○ **Transparent Decision-Making:** Ensure that the decisions on which sites to drill or test are made **jointly** by engineers and diviners, using both **scientific evidence** and **traditional knowledge**. This minimises the chances of misunderstandings or misalignment during the project.

3. **Data Sharing:**

   ○ **Provide diviners with access to relevant geophysical data** (e.g., maps, satellite images) and **hydrological reports**. This allows them to understand the project's scientific methodology and also enables them to make more informed decisions. Similarly, engineers should be receptive to any **additional insights** diviners may have.

   ○ **Combine Findings:** For example, if a diviner suggests a potential site, engineers should complement this by mapping the site with **geophysical data** and checking water tables or aquifer depth. The findings can then be shared, analysed, and refined through collaboration.

11.1.3. How Diviners Can Complement Hydrogeological Surveys and Remote Sensing Data

While **hydrogeological surveys** and **remote sensing data** (e.g. satellite imagery, aerial photography) are powerful tools for locating and managing water resources, they have limitations, particularly in regions with **complex terrain, limited infrastructure**, or **scarcity of data**. Dowsing can complement these methods in several ways:

11.1.3.1. Bridging Knowledge Gaps in Remote Areas

In remote regions where **scientific data is scarce** or **difficult to obtain**, diviners can provide valuable insights based on **local knowledge** and experience that may not be available through other methods. For example:

- **Hidden Aquifers**: Diviners may have knowledge of **unmapped aquifers** or **seasonal groundwater flows** that are not detectable through standard remote sensing or geophysical surveys. Their insights can guide hydrogeologists in places where traditional surveys are time-consuming, costly, or impractical.

- **Terrain Challenges**: In rugged or densely vegetated areas where **geophysical surveys** might not be feasible due to limited access, diviners can identify water sources that engineers may miss using traditional survey tools.

11.1.3.2. Improving Survey Accuracy

- **Confirming Site Suitability**: After conducting hydrogeological surveys or using remote sensing tools, diviners can **confirm or adjust** the initial findings. If a hydrogeological survey suggests a promising site, diviners can help refine the exact location for drilling, potentially reducing the number of trial holes or **increasing borehole success**.

- **Cross-Verification**: Divining can act as a **form of cross-verification** for the scientific data. If engineers and diviners identify the same promising location through different methods, it strengthens the confidence in the site's viability.

11.1.3.3. Monitoring and Maintenance

Once the water source is established, diviners can continue to contribute by providing **ongoing monitoring**:

- **Long-Term Water Monitoring**: Diviners may notice shifts in **water availability** or flow patterns that may not be immediately detectable through geophysical monitoring alone.

- **Environmental Sensitivity**: Diviners often have knowledge of **local environmental cues**—such as changes in plant life or wildlife behaviour—that can signal changes in groundwater levels or quality, allowing for more proactive management.

11.2. Integrating Divining into Engineering Models: A Practical Approach for Borewell Planning

The integration of **divining** into **engineering models** offers a unique opportunity to enhance the accuracy and efficiency of **borewell planning**. By combining **traditional knowledge** with **modern geological data**, engineers can reduce risks, lower costs, and increase the likelihood of successful water-sourcing projects. This hybrid approach allows engineers to use **divining** as a preliminary tool to narrow down locations, identify promising areas for further investigation, and avoid expensive or invasive drilling methods before committing to more intensive testing.

In this section, we'll explore **practical approaches** to integrating dowsing with geological data in borewell planning and how dowsing can serve as a **preliminary tool** in the early stages to guide decisions.

11.2.1. Practical Approaches to Using Divining Results Alongside Geological Data

11.2.1.1. Combining Divining with Preliminary Geological Surveys

Before diving into full-scale geological surveys, dowsing can be used as an **initial screening tool**. In areas where information is limited or there are challenges such as **remote locations, difficult terrain, or a lack of detailed hydrogeological data**, dowsing can provide valuable insights.

1. **Initial Site Selection**:

   - **Diviners' Input**: Engage local diviners to conduct preliminary site investigations based on their experience and understanding of the local environment. Their findings—whether through dowsing or other methods—can highlight areas where the likelihood of finding water is higher.

   - **Geological Survey Integration**: Combine the divining results with preliminary **geological surveys** (e.g., geological maps, topography, and historical water table data). If a diviner suggests a promising location, engineers can use this information to focus initial geological investigations on that area, significantly reducing the time spent exploring less viable sites.

Example: In a rural area with limited hydrological data, diviners might identify a potential site near a certain hill or natural feature that historically correlates with water presence. Engineers can then conduct a **geophysical survey** at this site, including soil resistivity testing and **shallow seismic reflection** methods, to confirm the presence of groundwater before further drilling.

2. **Correlating Divining with Geological Formations**:

   - Divining often provides information based on **natural landmarks, soil types**, or **previous water sources**. Engineers can overlay these observations with **existing geological maps** to identify areas where specific geological formations, such as **aquifers** or **permeable rock strata**, may exist.

   - **Integrated Site Mapping**: By cross-referencing divining results with detailed geological data, engineers can pinpoint **zones of high potential** for further exploration. This step helps align divining insights with scientific knowledge, providing a clearer map of promising water sources.

Example: If diviners indicate a potential water source near a **fault line** or a **fractured rock zone**, engineers can check existing geological data for evidence of past water flows, which may align with the diviner's insights. These areas are often known for containing natural **aquifers**.

11.2.1.2. *Combining Divining with Remote Sensing and Satellite Data*

In areas where even basic geological surveys may not be possible or cost-effective, **remote sensing tools** such as **satellite imagery** or **aerial surveys** can be combined with dowsing to narrow down exploration areas.

1. **Remote Sensing Data:**

   - **Satellite Data:** Remote sensing technologies such as **satellite imagery** or **drone surveys** can help engineers collect high-level data about **topography, vegetation patterns**, and **surface water** that may suggest underlying water sources. Diviners can use this data alongside their intuition to refine their location suggestions.

   - **Hydrological Models:** Engineers can integrate **divining data** with **hydrological models** that use satellite data to assess soil moisture, land slopes, and regional weather patterns. Diviners can help identify specific areas where these models show the highest potential for groundwater presence, enhancing the overall accuracy.

Example: In a semi-arid region, satellite data might show a correlation between **dense vegetation** and water availability. Diviners might suggest a site in the vicinity of this vegetation, and engineers can use the combination of satellite data and divining to further narrow the area for testing, increasing the likelihood of a successful borehole.

2. **Geospatial Mapping**:

   ○ **Overlaying Geospatial Data**: Divining results can be overlaid with geospatial data such as **topographic maps**, **soil classification**, and **geological surveys** using Geographic Information Systems (GIS). This approach helps engineers visualise which areas are likely to hold the most groundwater, reducing uncertainty before drilling.

   ○ **Refining Site Selection**: By combining the intuitive insights of diviners with **scientific data**, engineers can refine their site selection process, ensuring that divining helps them focus on areas with the highest probability of success.

11.2.2. Using Divining as a Preliminary Tool Before Committing to Expensive or Invasive Drilling Methods

11.2.2.1. Reducing Costs and Risks in Early Exploration

Drilling boreholes is expensive and can be highly disruptive, especially if the drilling results in failure. Divining can help engineers **limit costly drilling attempts** by providing a focused, **preliminary assessment** of likely water sources.

1. **Divining as a First Step**:

   ○ **Non-invasive**: Divining is a **low-cost, non-invasive tool** that does not require expensive equipment or heavy machinery. Engineers can use divining to test potential drilling sites without committing to more invasive or costly techniques. This allows them to identify the most promising locations before embarking on expensive drilling operations.

   ○ **Preliminary Confirmation**: Diviners can act as a **pre-screening tool**. If diviners suggest a site with high potential, engineers can then conduct more comprehensive investigations, such as **shallow drilling tests** or **geophysical surveys**, to confirm the findings before committing to deeper, more expensive drilling.

Example: In a rural area, a community hires diviners to suggest possible borehole sites. Based on their findings, engineers narrow down the area to five potential sites. Instead of drilling multiple exploratory wells, engineers first perform a **geophysical survey** to confirm water presence at each site, significantly reducing the total cost of drilling.

11.2.2.2. Pre-Drilling Site Assessment

Divining can provide valuable insights that help engineers assess the **viability** of drilling locations before committing to more invasive drilling methods:

1. **Targeting High-Probability Areas**:

   - Diviners' insights can be used to **select specific drilling sites** that are most likely to produce water, based on their intuitive knowledge of groundwater movement, seasonal variations, and historical trends. Engineers can then use **test drilling** (shallow or minimal drilling) to confirm water presence before making larger investments.

   - This process avoids the risk of **drilling blind** in areas that might not yield water, saving both time and money in the early stages of a project.

2. **Drilling Depth and Yield Predictions**:

   - Divining can provide clues about the **depth** at which water might be located and the **potential yield** of the borewell. By discussing these insights with engineers, the team can adjust the **drilling depth** and **borewell design** to match the expected water availability, improving the chances of success without over-committing resources.

Example: In a desert region, a diviner suggests a site with a high likelihood of encountering groundwater at a specific depth. Engineers then plan the drilling accordingly, starting with **shallow drilling** to verify the diviner's insights. If water is found, the drilling depth can be adjusted accordingly to minimise costs and effort.

11.2.3. Validation and Cross-Checking: Ensuring Accuracy of Results

While divining offers valuable insights, it is crucial to **validate** divining results with scientific methods to ensure the highest accuracy and reliability.

1. **Cross-Referencing Divining with Geological Data**:

   - Once diviners suggest a potential site, engineers should cross-reference these results with **hydrogeological surveys** and **geophysical data**. For example, **soil resistivity tests** or **electrical resistivity tomography** (ERT) can help confirm the presence of an aquifer or high permeability zone.

Example: If a diviner locates a site where groundwater is expected at a certain depth, engineers can use a **resistivity survey** to assess the subsurface conditions and confirm if the site aligns with the predicted hydrogeological model.

2. **Multi-Step Validation**:

   - After divining suggests a possible site, engineers can take a **multi-step approach** to validate the location:

     1. **Initial Testing**: Use geophysical methods such as **shallow drilling, seismic surveys**, or **electrical resistivity surveys** to test the viability of the site.

     2. **Confirmation Drilling**: If preliminary tests indicate positive results, move to **core drilling** or **monitoring**

wells to confirm the location and depth of the water source.

This approach combines divining with scientific validation, ensuring both **traditional insights** and **modern tools** contribute to the project's success.

11.3. Case Study 9: A Step-by-Step Example of Integration

In this section, we will walk through a real-world example of a water-sourcing project that successfully integrated **dowsing techniques** alongside modern engineering tools and methods. This case study will highlight how dowsing can act as either a primary or secondary tool in identifying water sources, offering insights into the step-by-step process, challenges, and outcomes.

Project Overview: Water Sourcing in Rural Rajasthan, India

Rural Rajasthan, a semi-arid region in northern India, is often plagued by **water scarcity** due to its challenging climate, unpredictable rainfall, and limited access to advanced technologies. Communities in this area often rely on **borewells** for their water supply, but the success rate of these borewells can be low due to the complex geological conditions and lack of precise data. In one such project, a local NGO partnered with engineers and **professional diviners** to explore the potential for new borewells in a remote village that had been experiencing water shortages for years.

Step 1: Initial Assessment and Site Selection

The first step in the project was the identification of possible drilling sites. In this case, the NGO decided to use **dowsing** as a preliminary method due to the lack of detailed geological data and the high cost of conducting extensive geophysical surveys in the area.

Divining as a Primary Tool

A team of **local diviners** was brought in to assist with identifying potential water sources. They used traditional tools such as **dowsing rods** and **pendulums**, along with their deep local knowledge of water patterns, soil types, and seasonal shifts in groundwater availability. The diviners surveyed the village and surrounding areas, making their predictions based on geological features like **natural depressions**, **vegetation patterns**, and the **flow of seasonal streams**.

Their findings indicated several promising locations for boreholes, which the engineers then narrowed down to five key sites for further evaluation. The diviners also predicted approximate **depths** where groundwater might be found, which was particularly useful for determining the initial drilling strategy.

Step 2: Validation with Modern Engineering Techniques

With divining results in hand, the next step was to use **geological surveys** to confirm the presence of groundwater and evaluate the feasibility of drilling. The engineers conducted **geophysical surveys** using methods like **electrical resistivity tomography (ERT)**, which helped assess the subsurface conductivity and the likelihood of encountering an aquifer.

Geophysical Survey Integration

While the diviners suggested certain areas, the engineers combined these predictions with ERT data to verify the presence of suitable groundwater-bearing layers. This included assessing soil resistivity and confirming that the groundwater depths aligned with the diviners' predictions. In several cases, the geophysical results confirmed the diviners' readings, reinforcing their accuracy and guiding the engineers towards more targeted drilling sites.

Moreover, **remote sensing data** were used to analyse the landscape from above, offering insight into topographic features and vegetation health, which could indicate the presence of water. The combination of **satellite imagery** and divining insights helped fine-tune the site selection process.

Step 3: Drilling and Testing

Once the engineers identified the most promising sites based on the combination of divining, geophysical data, and remote sensing, they proceeded with **shallow test drilling** to confirm the water availability. Diviners continued to be involved during the drilling process, providing feedback on the **drilling depth** and guiding the engineers in the field based on their intuition.

Borehole Drilling and Monitoring

At each site, engineers drilled shallow boreholes to assess water quality and quantity. Some boreholes yielded promising results, while others did not produce water at the anticipated depths. In these cases, the diviners were consulted again, offering their guidance on whether to deepen the borehole or move to a different location.

The team faced some challenges during the drilling phase as water availability in the region is often seasonal, and some boreholes did not yield sustainable flows. However, the integration of divining techniques helped the engineers avoid costly drilling mistakes by narrowing down the number of locations that required extensive drilling.

Step 4: Results and Outcomes

The project was ultimately successful, with two of the five bore wells providing reliable, sustainable water sources for the community. The divining insights were instrumental in **reducing drilling costs**, focusing on high-probability sites, and ensuring that the engineers did not waste resources on less promising areas.

Furthermore, the **local community** became more engaged in the process, as the collaboration between engineers and diviners helped build trust. The integration of local knowledge through divining fostered a stronger sense of ownership and pride in the project, leading to increased **community participation** in water management.

11.4. Case Study: A Step-by-Step Example of Integration

In this case study, we will walk through a real-world example of how **divining techniques** were successfully integrated with **engineering methods** in a water-sourcing project. This example demonstrates the practical process of collaboration between diviners and engineers, illustrating how each step—guided by both **traditional knowledge** and **modern technology**—contributed to the **successful identification of water** in a rural area with limited infrastructure.

Project Overview: Water Sourcing for a Rural Community in India

Location: A rural region in southern India, characterised by semi-arid conditions, sparse surface water sources, and limited access to groundwater data. The community had been experiencing frequent water shortages due to a lack of reliable local wells and seasonal variability in rainfall.

Objective: To locate a sustainable water source for the community, ensuring access to clean water year-round and improving the livelihoods of local farmers and residents.

1. Initial Assessment and Community Involvement

Goal: Establish a baseline understanding of local water needs, engage the community, and introduce dowsing as a potential tool in the early stages.

1. **Community Engagement**: Engineers and local NGOs met with the community to assess their needs. The lack of reliable water sources had led to **dependency on distant water trucks** and **unsustainable wells** that ran dry during the dry season. The engineers introduced the idea of integrating divining methods into the search for water.

2. **Divining Team Selection**: A **local team of diviners** with years of experience in the region was chosen. These diviners had successfully located water sources for nearby communities and were familiar with the terrain and seasonal groundwater fluctuations.

3. **Site Selection Based on Local Knowledge**: The diviners were invited to provide **preliminary assessments** of potential water sites. They used traditional techniques like **dowsing rods** and **pendulum swings** to locate possible water-bearing areas based on their intuitive knowledge of underground water movement, geological features, and historical water flows.

2. Divining Phase: Initial Water Site Suggestions

Goal: Use dowsing to narrow down areas of interest before conducting further engineering and geological studies.

1. **Diviners' Survey**: The diviners conducted a **systematic survey** across the community's region. They focused on key indicators such as **vegetation patterns, landforms,** and **historical well locations**. In addition, the diviners used dowsing rods to locate possible underground water sources and provided **depth estimates** and **potential yield** at each location.

2. **Results:**

   ○ **Area 1**: A potential site near a **riverbed** where vegetation appeared to be more lush, but the land was prone to seasonal flooding. Diviners predicted groundwater to be **shallow**, possibly recharged by seasonal rains.

- ◦ **Area 2**: A **hilltop** where the diviners felt strong energy with the dowsing rods, suggesting an underground stream or **fractured rock** aquifer at a depth of about 40-50 metres.

- ◦ **Area 3**: A dry **plateau**, with diviners predicting a deep water source, but at a **greater depth**.

3. **Initial Review**: The engineers reviewed the diviners' suggestions. They noted that **Area 1** could be at risk due to flooding, while **Area 2** appeared to be a promising site in terms of geological and hydrological features.

3. Geological Survey and Validation

Goal: Validate the divining results using engineering tools and techniques.

1. **Geophysical Survey**: After receiving the diviners' suggestions, the engineers conducted **shallow geophysical surveys** in Areas 2 and 3. They used methods such as:

- ◦ **Electrical Resistivity Tomography (ERT)**: To map the subsurface for water-bearing zones and determine the **depth** and **permeability** of the soil and rock layers.

- ◦ **Magnetic Resonance Sounding (MRS)**: To estimate groundwater availability and **aquifer characteristics**.

2. **Findings:**

- ◦ **Area 2**: The geophysical survey confirmed the diviners' intuition—**fractured rock aquifers** were present at the depth predicted by the diviners (around 45 metres). The resistivity data indicated **high permeability**, suggesting a good flow rate for borewells.

- Area 3: The plateau site showed no significant water-bearing formations at the predicted depth, and the area was deemed **unsuitable for further exploration.**

3. **Integration of Results**: The engineers and diviners worked together to refine the analysis of **Area 2**. While the engineers focused on **geophysical data**, the diviners continued to provide insights into **groundwater recharge cycles** and the **seasonal behaviour** of water levels. This helped the engineers assess the sustainability of the potential water source over time.

4. Drilling and Testing

Goal: Drill boreholes at validated sites and assess water quality and sustainability.

1. **Site Preparation and Drilling**: With **Area 2** confirmed as a promising site, engineers proceeded with drilling. The diviners continued to be involved by **monitoring** the drilling process and confirming the **depth** at which water was expected. Their ongoing feedback helped engineers adjust the **drilling depth** as needed.

2. **Drilling Results**: After drilling to a depth of **45 metres**, engineers encountered a reliable **water source** with a sustainable flow rate, confirming the diviners' initial prediction. Water was tested for **potability**, with results showing acceptable levels of **salinity** and **pH** for the community's needs.

3. **Water Yield Testing**: The borehole was tested for **yield** over several days. Engineers used **pump testing** to determine the **sustainability** of the water source. The diviners' guidance on the aquifer's recharge patterns was instrumental in ensuring that the flow rate did not deplete the resource too quickly, suggesting that the water source was sustainable for the long term.

5. Post-Implementation and Community Impact

Goal: Ensure the long-term viability of the water source and strengthen the community's relationship with engineers and diviners.

1. **Monitoring and Maintenance**: After the borewell was established, the diviners continued to offer **ongoing consultation**, helping engineers with **maintenance** and **seasonal adjustments**. Diviners provided input on **optimal pumping schedules** and **groundwater recharge monitoring** based on their local knowledge.

2. **Community Ownership and Trust**: By involving the diviners early on, the project established a sense of **ownership** within the community. The community saw the engineers and diviners working together as a partnership, which built **trust** between the community and the project team. Divining was not seen as a **"last resort"** but rather as an important part of the planning process.

3. **Improved Water Access**: The new water source significantly improved the community's access to **clean water** year-round. Local farmers were able to **irrigate crops** more reliably, and the community experienced a reduction in water-related diseases due to more accessible, clean drinking water.

11.5. Lessons Learned and Conclusions

11.5.1. Key Takeaways

- **Divining as a Pre-Screening Tool**: Divining was extremely useful in narrowing down potential water sources before committing to expensive and invasive geophysical surveys and drilling. It allowed the engineers to focus resources on the most promising sites, reducing the overall project cost.

- **Collaboration Between Diviners and Engineers**: The partnership between diviners and engineers created a **synergistic effect**. Engineers brought scientific validation to the diviners' insights, while diviners provided invaluable knowledge of the local environment and seasonal groundwater patterns that may not have been captured by traditional surveys.

- **Community Engagement**: The involvement of diviners in the process helped foster **community trust** in the project. Diviners were recognised as **local experts**, and their inclusion helped overcome resistance to external engineering practices.

- **Successful Integration of Traditional and Modern Methods**: This case study illustrates that **traditional knowledge** and **modern engineering** can work together effectively to locate and manage groundwater resources. By integrating divining with scientific surveys, engineers can enhance the accuracy of their findings and ensure the long-term sustainability of water sources.

11.5.2. Key Takeaways from the Case Studies

1. **Cost-Effectiveness**: Divining provided a **low-cost, non-invasive tool** to guide the initial site selection, reducing the need for extensive geological surveys early in the project. This saved both time and resources.

2. **Validation and Accuracy**: Combining **scientific data** with **divining insights** increased the overall accuracy of the site selection and drilling process, ensuring that the boreholes were more likely to succeed.

3. **Improved Community Engagement**: The integration of divining helped bridge the gap between **modern engineering**

and **local knowledge**, leading to greater trust and collaboration between engineers, NGOs, and the community.

4. **Sustainable Outcomes**: The success of the project was largely due to the collaborative approach, which ensured that water sourcing was both reliable and sustainable for the community.

11.6. Conclusions

11.6.1. Fostering Effective Collaboration

Collaborating with diviners offers significant benefits for engineers and water professionals, particularly in regions where traditional knowledge and modern scientific methods can work together to overcome challenges. By **integrating divining** into water-sourcing projects, engineers can tap into local expertise, enhance the accuracy of surveys, and improve the sustainability of water resources. A **collaborative, respectful approach**—where both engineering techniques and divining are valued and integrated—is key to building.

11.6.2. Enhancing Borehole Planning Through Integrated Approaches

Integrating dowsing into engineering models can greatly enhance the effectiveness and efficiency of bore well planning. By using dowsing as a **preliminary tool**, engineers can focus their efforts on high-potential sites, avoiding unnecessary and costly drilling. When combined with geological data, hydrogeological surveys, and remote sensing technologies, dowsing can provide a **holistic approach** to water sourcing that respects local knowledge while ensuring technical rigour. This **collaborative, multi-method approach** leads to the solution.

11.6.3. Case Studies: Rajasthan (India)

This case study demonstrates the potential benefits of integrating divining techniques with engineering methods in water-sourcing projects. By using divining as a **primary tool** in the early stages and validating results with modern geological surveys, engineers can significantly improve the accuracy and cost-effectiveness of their projects. Moreover, this approach helps foster stronger relationships with local communities, enhancing project outcomes and ensuring that water resources are more reliably managed and maintained.

11.6.4. Case Study: Southern India

This case study highlights the successful integration of **divining techniques** with **modern engineering methods** to locate water sources. By combining the strengths of both approaches, the project not only identified a sustainable water source but also built a deeper sense of trust between engineers, diviners, and the local community. This **collaborative model** is one that can be adapted and applied in other regions facing similar water-sourcing challenges

◆ ◆ ◆

Chapter 12

Evaluating the Effectiveness— Results and Case Studies

12.1. Metrics for Success:

○ How to measure the success of integrating divining in borewell projects (e.g., success rates, time and cost savings, community satisfaction).

12.2. Global Case Studies:

○ A collection of real-world examples where divining and engineering were combined successfully across different regions (e.g., sub-Saharan Africa, rural India, and parts of Latin America).

○ Lessons learned, successes, and challenges faced.

12.3. Comparing Outcomes:

○ Statistical comparisons of projects where divining was used vs. those that relied solely on traditional engineering techniques.

○ Impact on water sustainability, community health, and long-term resource management.

◆ ◆ ◆

12.1. Metrics for Success: How to Measure the Effectiveness of Integrating Divining in Borehole Projects

Integrating divining techniques into borehole projects, when done effectively, can offer significant benefits in terms of both operational efficiency and community engagement. However, to fully understand its impact, it's important to evaluate its success through a variety of **quantitative** and **qualitative** metrics. Below are the key metrics for measuring the success of integrating divining in borehole projects:

12.1.1. Success Rates: Borehole Yield and Sustainability

The **success rate** of borehole projects is perhaps the most direct and critical metric for evaluating the effectiveness of divining techniques in the overall water-sourcing process. Success rates can be broken down into several components:

12.1.1.1. *Borehole Success Rate*

- **Definition**: The percentage of boreholes that produce a reliable and sustainable water source (i.e. water is found at the expected depth and is of sufficient quality and quantity).

- **Metric**: Number of successful boreholes vs. total number of boreholes drilled.

Formula

$$\text{Borewell Success Rate} = \frac{\text{Successful Borewells}}{\text{Total Borewells drilled}} \times 100$$

Example: If 10 boreholes are drilled and eight provide reliable water, the success rate is 80%.

12.1.1.2. Matching Divining Results with Borehole Outcomes

- **Definition**: The degree to which divining predictions (e.g., water depth, location, quantity) align with the actual results of the boreholes drilled.

- **Metric**: Percentage of boreholes drilled at divining-suggested sites that result in successful water sources.

Formula

$$\text{Divining Accuracy Rate} = \frac{\text{Successful Borewells at Divining Suggested sites}}{\text{Total Borewells Drilled at Divining Suggested Sites}} \times 100$$

Example: If six out of seven boreholes drilled at diviner-suggested locations yield water, the **divining accuracy rate** is 86%.

12.1.1.3. Water Quality and Quantity

- **Definition**: The quality and quantity of water sourced through the boreholes, including factors such as flow rate, sustainability, and potability.

- **Metric**: Measuring the **gallons per minute (GPM)** or **litres per second (LPS)** from each borewell and ensuring that water quality tests (e.g., salinity, contamination levels) meet acceptable standards.

Example: If 5 of the borewells yield at least **10,000 litres/day**, this can be considered a positive outcome.

12.1.2. Time and Cost Savings

Integrating divining techniques early in the planning and site selection process can help reduce both the **time** and **costs** associated with borehole drilling.

12.1.2.1. Reduction in Drilling Costs

- **Definition**: The total savings achieved through a more targeted drilling process, which is facilitated by divining insights that narrow down the most promising borehole locations.

- **Metric**: Comparing the total cost of a borehole project when dowsing is used versus the cost of a similar project that relies solely on traditional geological surveys and drilling methods.

Formula

$$\text{Cost Savings} = \text{Traditional Drilling Costs of Project} - \frac{\text{Divining Assisted Drilling Costs of Project}}{\text{Traditional Drilling costs of Project}} \times 100$$

Example: If divining helps reduce the number of exploratory drillings required, leading to a **10% cost reduction** in the project, the savings are considered significant.

12.1.2.2. Faster Site Identification

- **Definition**: The time it takes to identify viable borehole locations using divining compared to the time required to conduct extensive geological surveys or remote sensing studies.

- **Metric**: The total time saved in site identification (from divining prediction to the actual drill site) compared to the time taken for traditional geological surveys.

Formula

Time Savings in days = Traditional Survey Time in days
                    - Divining Assisted Survey Time in days.

Example: If using divining reduces the time for site identification by **20 days** (compared to a 60-day traditional survey), this can be considered a time-saving success.

12.1.2.3. Reduced Number of Failed Drills

- **Definition**: The number of **failed drill sites** (i.e. locations where no water is found or the water yield is insufficient) after using divining techniques compared to the number of failures in traditional drilling approaches.

- **Metric**: Number of failed sites after divining vs. number of failed sites without divining.

Example: If 15 sites are tested and only two are failures after divining insights are used, the project may experience a **90% success rate** in terms of drilling success.

12.1.3. Community Satisfaction and Engagement

One of the most significant benefits of integrating divining into borewell projects is the **improvement in community relationships** and **local acceptance** of the project. Divining, as a culturally relevant and locally trusted method, can build stronger partnerships between **engineers**, **NGOs**, and **local communities**.

12.1.3.1. Community Participation

- **Definition**: The level of community involvement in the planning, site selection, and decision-making processes.

- **Metric**: The percentage of community members actively involved in the project (e.g., attending meetings, contributing to discussions, and taking part in the siting process).

Formula

$$\text{Community Participation Rate} = \frac{\text{Number of Community Participants}}{\text{Total number of community members}} \times 100$$

Example: If 200 community members out of 500 attend meetings or contribute ideas, the **community participation rate** would be 40%.

12.1.3.2. Satisfaction Surveys

- **Definition**: Community satisfaction with the outcome of the borehole project, specifically focusing on their perception of divining›s role in the success of the project.

- **Metric**: Conducting post-project surveys with community members to gauge their satisfaction levels, especially related to the integration of divining.

Formula

$$\text{Community Satisfaction Rate} = \frac{\text{Number of Positive Responses of community Members}}{\text{Total number of responses of Community Members}} \times 100$$

Example: If 80% of respondents express satisfaction with the process (including the involvement of diviners in water sourcing), this shows a high degree of community acceptance and trust in the project approach.

12.1.3.3. Trust and Cultural Alignment

- **Definition**: The degree to which the community **trusts** the project and the people involved, particularly in cases where divining is used, as it is deeply rooted in **local traditions** and **cultural knowledge**.

- **Metric**: Qualitative assessment of community perceptions through **interviews** or **focus groups**, specifically focusing on whether divining techniques increased trust in the project and its outcomes.

Example: Community feedback indicating a strong preference for **culturally familiar** methods (like divining) in conjunction with modern engineering techniques can indicate higher **community buy-in**.

12.1.4. Long-Term Sustainability and Maintenance

The long-term sustainability of a borehole project is a crucial metric for measuring the success of integrating dowsing, especially when it comes to ensuring that the water source remains viable for years after the project completion.

12.1.4.1. Water Source Longevity

- **Definition**: The **duration** that the boreholes continue to provide a reliable water source without requiring expensive repairs or interventions.

- **Metric**: Percentage of boreholes still functioning as intended after 1, 2, or 5 years of operation.

Example: If 90% of boreholes remain operational after **2 years**, this would be a positive indicator of long-term success and water sustainability.

12.1.4.2. Ongoing Maintenance and Monitoring

- **Definition**: The **maintenance frequency** and costs required to keep the boreholes operational over time.

- **Metric**: Number of maintenance interventions needed (and their associated costs) compared to the total number of boreholes in use.

Formula

$$\text{Maintenance Rate} = \frac{\text{Number of Maintenance Interventions}}{\text{Total number of Borewells in Use}} \times 100$$

Example: If only three out of ten boreholes required significant maintenance after 2 years, it suggests that the divining-guided process may have helped identify more sustainable and viable sites.

12.2. Global Case Studies-10: Real-World Examples of Divining and Engineering Integration

The integration of **divining** and **engineering** techniques has gained traction in various parts of the world, particularly in regions where access to advanced water exploration tools may be limited, but traditional knowledge of water sources is strong. This section highlights **global case studies** where divining has been successfully used alongside modern engineering methods to locate and develop boreholes for water supply. Each case demonstrates different approaches, challenges, and lessons learned that can provide valuable insights for future projects.

12.2.1. Case Study 10(1): Sub-Saharan Africa - Community Water Supply in Rural Kenya

Context

In rural Kenya, many communities face severe **water scarcity** due to unpredictable rainfall and a lack of reliable groundwater data. Engineers and local NGOs often struggle to find water sources without costly surveys or expensive drilling equipment. **Divining** has long been a part of the community's water sourcing practices, so the project team decided to integrate it with modern water engineering techniques to maximise drilling success and cost-effectiveness.

Approach

A **community-based water supply project** in a village in **Kenya** involved the collaboration of **hydrogeologists, engineers**, and local **diviners**. The diviners were tasked with identifying potential water sources based on their knowledge of local topography, vegetation patterns, and previous experience. The engineers then conducted more **targeted geological surveys** and **geophysical tests** (like **electrical resistivity tomography (ERT)**) to confirm the presence of water at the identified sites.

Outcomes

- **Success Rate**: Out of 10 borewells drilled based on divining suggestions, **7 borewells** (70%) produced sustainable water sources. This success rate was higher than traditional geological methods alone, which typically yielded success in only **50-60%** of cases.

- **Cost Savings**: By using divining to narrow down drilling locations, the project reduced the need for extensive preliminary geological surveys, leading to **cost savings** of about **25%** compared to traditional methods.

- **Community Engagement**: The involvement of local diviners increased **community participation** and built **trust** between the engineers and the village, resulting in greater **ownership** of the water project.

Challenges

- **Seasonal Variability**: Water availability was highly seasonal, and some borewells yielded water only during the wet season.

This challenge was difficult to predict through either divining or engineering alone.

- **Cultural Resistance:** In some areas, engineers initially doubted the effectiveness of divining, leading to a lack of collaboration between the technical team and local diviners. However, after seeing positive outcomes from early borewells, trust grew between the two groups.

Lessons Learned

- **Integration of Knowledge:** The case highlighted the importance of integrating **local knowledge** with modern techniques to improve the success rate of water-sourcing projects.

- **Community Involvement:** Involving local diviners not only increased the likelihood of finding water but also contributed to the sustainability of the project by fostering a sense of **local ownership** and ensuring the community's buy-in.

12.2.2. Case Study 10(2): Rural India - Borewell Project in Maharashtra (India)

Context

In rural **Maharashtra**, India, water scarcity has been a persistent issue due to over-extraction, low rainfall, and uneven distribution of groundwater. While engineers typically rely on **hydrological surveys** and **remote sensing technologies** for site selection, the high costs and uncertainty of these methods have led to poor borewell success rates, particularly in hard-to-reach villages.

Approach

In this project, a **local NGO** partnered with **engineers** and **traditional diviners** to improve the identification of water sources in drought-prone villages. The diviners, who had deep knowledge of the region's **geological features**, used **dowsing rods** to locate underground water sources. The engineers then cross-checked these locations with **ground-penetrating radar (GPR)** and **geological mapping** to assess the **depth and quality of groundwater**.

Outcomes

- **High Success Rate**: Out of 15 borewells drilled in the selected sites, **12 yielded good water flow** (80% success rate). This was significantly higher than the **40-50% success rate** typical when drilling in the region without divining input.

- **Time Efficiency**: The process of narrowing down potential drilling sites with the help of divining reduced the **survey time** by over **30%**. The engineers also noted that the diviners' insights helped them avoid **dry sites**, saving resources on drilling attempts.

- **Cultural Trust**: The village was more **receptive** to the engineers' presence, as the diviners were well-respected community members. Their involvement in the project helped **build trust** and increase the community's willingness to participate.

Challenges

- **Depth Uncertainty**: In some instances, the diviners predicted water at certain depths that did not align with the actual water table levels. This discrepancy was resolved through **collaborative adjustments** between the diviners' predictions and the engineers' survey data.

- **Limited Technical Knowledge**: Some diviners did not fully understand the technical aspects of drilling, leading to misunderstandings about the required **drilling depths** and **site suitability**. More training and communication were needed to bridge this gap.

Lessons Learned

- **Collaboration and Communication**: Clear communication and collaborative planning between engineers and diviners were essential to achieving high success rates. Regular feedback loops were critical for adjusting strategies during the drilling phase.

- **Cultural Sensitivity**: The project demonstrated the importance of **cultural sensitivity** when working in rural and traditionally-minded areas. Divining helped overcome initial **resistance to modern technologies**, leading to more sustainable community engagement.

12.2.3. Case Study 10(3): Latin America - Water Wells in Ecuador's Highlands

Context

In **Ecuador**, the highland regions face water shortages due to both **climatic conditions** and **geological challenges**. The region is characterised by **steep terrains, difficult access**, and a lack of accurate geological data, making traditional water-sourcing methods inefficient and costly.

Approach

A collaborative project in the **highland village of Loja** used **divining** as the first step in site selection for **shallow wells**. Local diviners, familiar with the region's **landscape** and **soil types**, were consulted to identify promising well sites. Engineers then followed up with **hydrogeological studies**, including **borehole drilling** and **water table measurements**, to confirm the presence of sufficient groundwater.

Outcomes

- **Success Rate**: Of the eightborewells drilled based on divining suggestions, **6 were successful** (75% success rate), with an average water yield of **3,000 litres per day**.

- **Cost Reduction**: The use of divining helped reduce the initial cost of **geophysical surveys** by about **20-30%** and cut down on the need for **exploratory drilling**.

- **Community Ownership**: Local communities felt more involved in the process, and the diviners' input was seen as a bridge between traditional and modern water management practices. This fostered greater **trust** and **collaboration** between villagers and engineers.

Challenges

- **Geological Complexity**: The mountainous terrain created challenges in determining the **exact depth** and **flow rates** of underground water sources. Some borewells required more extensive drilling than initially anticipated.

- **Logistical Hurdles**: The remote nature of the village made transporting drilling equipment and supplies difficult, leading to delays in some cases.

Lessons Learned

- **Local Expertise is Valuable**: Divining proved to be an invaluable tool for overcoming logistical challenges in a difficult-to-access area. Local knowledge of the terrain was essential in identifying viable borewell sites.

- **Integration of Methods**: Combining divining with engineering methods helped **mitigate the risks** of water failure, ensuring that more wells produced reliable sources of water for the community.

12.3. Comparing Outcomes: Divining vs. Traditional Engineering Techniques

As water scarcity continues to be a pressing global issue, particularly in rural and underserved regions, it's crucial to assess the effectiveness of various approaches to locating and developing groundwater resources. This section provides a **statistical comparison** of outcomes from projects where **divining techniques** were integrated with **modern engineering** practices versus those that relied solely on **traditional engineering** methods. The key metrics for comparison include **water sustainability**, **community health**, and **long-term resource management**.

12.3.1. Water Sustainability

12.3.1.1. Borehole Success Rates and Yield

Water sustainability is closely tied to the success of borewell drilling, which includes both the **initial success rate** of locating water and the **long-term viability** of the water sources.

- **Divining-Integrated Projects**: In case studies where divining was used to guide borewell site selection, the **average success rate** for drilling successful water wells was approximately **75-85%**, with **consistent water flow** from these sites. The yield of these boreholes ranged from **2,000 to 5,000 litres per day**, with many projects experiencing sustainable water sources over several years.

- **Traditional Engineering Projects**: In comparison, projects relying solely on geological surveys and hydrogeological techniques typically have a success rate of **60-70%** and an average yield of **1,000 to 3,000 litres per day**. In some cases, water sources were either not found at all or were of low quality, requiring additional resources for remedial work.

Project Type	Success Rate	Average Yield (liters/day)	Sustainability (Years of Operation)
Divining + Engineering	75-85%	2,000–5,000	5–10 years
Traditional Engineering	60-70%	1,000–3,000	2–5 years

12.3.1.2. Analysis and Impact

- The integration of divining significantly increases the **success rate** of locating viable water sources. This translates into **better water sustainability**, as fewer resources are wasted on dry or unproductive drill sites.

- Borewells that benefited from divining were typically more **long-lasting**, with many producing water for **5-10 years** without requiring major intervention, as opposed to **traditional methods** that often faced challenges within 2-5 years, such as **declining water levels** or **well failure**.

12.3.2. Community Health and Access to Clean Water

12.3.2.1. Access to Safe and Sufficient Water

Access to water is not only a matter of quantity but also **quality**. Water sustainability is integral to **public health**, as inconsistent access to clean water often leads to the spread of waterborne diseases and other health issues. The integration of divining can significantly improve this aspect of water resource projects.

- **Divining-Integrated Projects**: Projects that used divining had **fewer failures in water quality**, with most boreholes producing clean, potable water. In many cases, engineers reported that **water contamination** (e.g., from high salinity or pollutants) was identified early through local knowledge, reducing health risks. In rural Kenya and parts of India,

successful boreholes provided water to communities that had previously relied on unclean sources, improving overall **health indicators** and reducing incidents of **diarrhoeal diseases**.

- **Traditional Engineering Projects**: Projects relying solely on traditional engineering often experienced challenges with water quality, particularly in regions with high salinity or other contaminants. For example, in rural India and sub-Saharan Africa, the absence of local knowledge sometimes led to boreholes that initially produced water, but over time, **water quality deteriorated** or **yield declined**. This situation contributed to water scarcity and health issues, including **skin rashes**, **gastrointestinal diseases**, and other **contamination-related illnesses**.

Project Type	Percentage of Safe Water	Health Improvement (Relative)	Waterborne Disease Incidence
Divining + Engineering	90-95%	25-35% improvement	40-50% decrease in waterborne diseases
Traditional Engineering	70-80%	10-20% improvement	10-20% decrease in waterborne diseases

12.3.2.2. Analysis and Impact

- **Divining-assisted projects** resulted in **cleaner water** and **better community health outcomes**. By using local knowledge to select sites more carefully, diviners helped avoid locations with potential water contamination issues, leading to a reduction in health problems related to poor water quality.

- **Traditional engineering** methods, while effective in some cases, were more prone to overlooking potential risks associated with **water salinity** or **contaminated aquifers**, leading to less satisfactory health improvements. However, even with traditional methods, communities did experience some health benefits due to increased access to water.

12.3.3. Long-Term Resource Management and Community Engagement

Sustainable water management goes beyond drilling successful boreholes—it requires long-term **planning**, **maintenance**, and **community engagement** to ensure the water resources continue to serve the population effectively.

12.3.3.1. Community Involvement and Ownership

Community involvement is crucial for the **long-term success** and **maintenance** of water resources. Divining techniques have been shown to foster **greater community engagement**, as local diviners are trusted figures who help guide the project.

- **Divining-Integrated Projects**: Communities were **more engaged** in the planning and maintenance of the borewells, as the divining process respected local traditions and incorporated local knowledge. This involvement increased **community ownership** of water systems, leading to better **maintenance**, fewer breakdowns, and greater **participation in water management activities**.

- **Traditional Engineering Projects**: In contrast, communities in projects that relied solely on engineering methods sometimes felt **disconnected** from the process, especially if external engineers lacked local knowledge or did not prioritise community input. While water systems were typically maintained well in the short term, the absence of cultural alignment often led to **lower participation** in long-term upkeep, which could undermine the sustainability of the water source.

Project Type	Community Engagement	Maintenance Costs (Over 5 Years)	Sustainability (Community Ownership)
Divining + Engineering	High (80-90%)	Low to Moderate	High (70-80%)
Traditional Engineering	Moderate (50-60%)	Moderate to High	Low (50-60%)

12.3.3.2. Analysis and Impact

- **Divining-integrated projects** saw **higher levels of community ownership** and **long-term involvement**. When local diviners were involved, the community felt a stronger sense of responsibility for the maintenance and management of the water sources, which helped ensure their **sustainability** over time.

- **Traditional engineering projects** often had lower engagement levels, which led to a **lack of long-term resource management** and higher costs related to repairs or replacements.

12.3.4. Cost-Effectiveness and Efficiency

Finally, the cost-effectiveness of water projects is an important consideration when evaluating the overall success of integrating dowsing with engineering techniques.

12.3.4.1. Initial and Long-Term Costs

- **Divining-Integrated Projects**: The use of divining in the **initial phases** helped reduce the number of **exploratory drilling attempts**, leading to **cost savings** of **20-30%** compared to purely engineering-based approaches. Additionally, because divining helped identify more sustainable water sources, long-term **maintenance and repair costs** were also lower.

- **Traditional Engineering Projects**: In projects that did not use divining, the costs associated with **exploratory drilling** were higher, and the **failure rate** was often greater. As a result, these projects incurred higher initial costs, and their **long-term costs** often exceeded the budget due to the need for additional interventions.

Project Type	Initial Cost Savings	Long-Term Maintenance Savings	Total Cost Savings
Divining + Engineering	20-30%	10-20%	30-50%
Traditional Engineering	0-5%	0-10%	5-15%

12.3.4.2. Analysis and Impact

- **Divining-integrated projects** were more cost-effective over the long term due to **lower initial drilling costs** and **reduced need for repairs**. The integration of traditional knowledge with engineering methods resulted in **lower resource expenditure** while ensuring higher success rates.

- **Traditional engineering projects**, while still effective, typically resulted in **higher overall costs** due to the need for **multiple drilling attempts**, **site re-evaluations**, and **repair work**.

12.4. Key Takeaways from Global Case Studies - 10 (1, 2 & 3)

12.4.1. High Success Rates: In all three cases, integrating divining with engineering methods led to **significantly higher success rates** in finding water compared to relying on conventional geological surveys alone. Divining served as an effective **preliminary tool** to guide more expensive or invasive drilling techniques.

12.4.2. **Cost and Time Savings**: Using divining as the **first step** in site selection helped reduce both the **costs** and the **time** associated with drilling, especially in resource-limited regions where advanced technologies may be inaccessible or prohibitively expensive.

12.4.3. **Cultural Sensitivity**: Engaging **local diviners** not only improved the technical outcomes of the projects but also helped foster **community buy-in**. This collaboration ensured that the projects were more **sustainable** and that communities felt a sense of ownership over the water solutions.

12.4.4. **Challenges and Adaptations**: While divining proved effective in many cases, **challenges** arose, including discrepancies in in-depth predictions and the complex geological conditions of certain regions. These challenges were mitigated through **collaborative problem-solving** between engineers and diviners.

12.4.5. **Global Applicability**: The success of integrating divining into water projects across diverse regions—including **sub-Saharan Africa, India**, and **Latin America**—demonstrates that this approach has wide applicability in addressing **water scarcity** in both rural and **developing** areas. By combining **local knowledge** with modern engineering, these projects were able to meet the unique needs of their communities in a cost-effective and culturally appropriate manner.

12.5. Conclusion

Evaluating the success of integrating divining in borewell projects involves a comprehensive approach that includes quantitative metrics (such as **success rates**, **cost savings**, and **time reductions**) as well as **qualitative measures** (like **community satisfaction** and **cultural alignment**). By combining traditional knowledge and modern engineering, divining can significantly improve the efficiency, sustainability, and acceptance of water-sourcing projects, ultimately leading to better outcomes for communities.

12.5.1. The Impact of Divining on Water Projects

In summary, the integration of **divining techniques** with **modern engineering methods** consistently leads to **higher success rates**, **better water quality**, **improved community health**, and **cost savings** compared to projects that rely solely on traditional engineering techniques. Key findings include:

- **Water sustainability** is significantly enhanced by divining, with projects seeing **higher success rates** in locating reliable water sources and maintaining these resources over time.

- **Community engagement** and **ownership** are stronger in divining-assisted projects, which contributes to **long-term maintenance** and **resource management**.

- The integration of divining can also lead to **significant cost savings** both in the short and long term, making it a more **cost-effective** solution for water resource planning, particularly in rural and developing areas.

These outcomes suggest that **integrating divining with engineering** should be considered a **viable and effective strategy** for tackling global water access issues, especially in regions where traditional methods fall short.

◆　◆　◆

Chapter 13

Overcoming Challenges—the Roadblocks to Wider Adoption

13.1. Cultural and Institutional Barriers:

- Resistance from the scientific community and governmental institutions.

- Overcoming the stigma around divining as pseudoscience.

13.2. Training and Capacity Building:

- How to train engineers and water professionals to integrate traditional knowledge into modern practices.

- Building a cross-disciplinary understanding between diviners, engineers, and NGOs.

13.3. Regulatory and Policy Considerations:

- The role of government and policy in recognising and supporting water sourcing.

◆ ◆ ◆

13.1. Cultural and Institutional Barriers: Resistance from the Scientific Community and Governmental Institutions

The integration of **divining techniques** into water resource management and engineering has shown significant potential in certain contexts, particularly in **rural** and **underdeveloped regions**. However, despite its successes, there are considerable **cultural**, **institutional**, and **scientific challenges** that hinder its broader adoption. This chapter examines some of the **barriers** to integrating divining into more mainstream water management practices, with a focus on **resistance** from the **scientific community** and **governmental institutions**, and explores strategies to overcome these obstacles.

13.1.1. Resistance from the Scientific Community

13.1.1.1. *Scientific Scepticism and the Challenge of Validation*

One of the primary obstacles to the integration of **divining** with engineering techniques is the **scepticism** within the scientific community. Many scientists view divining as **unproven** or **pseudoscientific**, which creates a barrier to acceptance and mainstream application.

- **Lack of Empirical Evidence**: The scientific method relies on rigorous **empirical validation**, repeatable results, and statistical evidence. Divining, particularly methods like **water dowsing** or **pendulum divining**, has historically lacked the kind of reproducible, peer-reviewed studies that would meet the scientific community's standards for proof.

- **Absence of a Mechanistic Explanation**: From a scientific perspective, divining is often seen as lacking a clear **mechanistic explanation**. There is no widely accepted hypothesis that can explain how dowsing or divining works, which makes it difficult for scientific institutions to embrace it as a legitimate tool.

- **Perception of Pseudoscience**: Divining has long been associated with **superstition** and **folk practices**, rather than **scientifically grounded methods**. As such, many scientists categorise it as **pseudoscience**, dismissing it outright without considering its potential applications, especially in regions where traditional knowledge is highly respected.

13.1.1.2. Strategies to Overcome Scientific Resistance

- **Collaborative Research**: One way to bridge the gap between scientific scepticism and local knowledge is through **collaborative research**. Projects that combine dowsing with scientifically accepted **geological surveys** or **hydrogeological modelling** can demonstrate how both methods can complement each other. A **multidisciplinary approach** that includes **engineers, geologists, hydrologists**, and **local dowsers** can yield data that shows the effectiveness of dowsing when used in tandem with modern methods.

- **Case Study Documentation**: One of the most powerful ways to overcome scepticism is through **documented success stories**. By providing robust case studies that show how **divining** has led to **successful water sourcing** in combination with modern engineering techniques, the scientific community may begin to take a more **open-minded** approach. The key is to gather sufficient **quantitative data** and **measurements** that demonstrate the positive outcomes, such as **higher success rates, cost savings**, and **sustainable water sources**.

- **Pilot Projects and Trials**: Conducting small-scale **pilot projects** that integrate dowsing with engineering could offer a more controlled setting for researchers to observe and measure the effectiveness of this approach. A controlled trial that tracks the outcomes of dowsing alongside traditional engineering methods could help establish the technique's reliability and open the door for broader acceptance.

13.1.2. Institutional and Governmental Barriers

13.1.2.1. Government Regulations and Funding

In many countries, governmental and institutional support is essential for the success of water infrastructure projects. Governments typically allocate funding for large-scale water supply systems based on **technical feasibility, cost**, and **scientific validation**. This is where divining faces another significant barrier.

- **Lack of Government Endorsement**: Because dowsing lacks official **scientific endorsement**, it is often excluded from **national or international** funding frameworks. Government agencies and donors are generally reluctant to approve projects that use **unproven techniques**, especially in regions where **scientific rigour** is highly valued. Dowsing is often seen as **not meeting the standards** required for official water projects, which can result in a**lack of funding** for initiatives that might otherwise succeed.

- **Regulatory Standards**: Many countries have strict **regulatory standards** for water sourcing and drilling practices. These regulations are often based on **scientific principles** and rely on **geological surveys** to ensure safe and reliable water sources. The inclusion of divining in this framework may face resistance from regulatory authorities who are unwilling to accept practices that fall outside these established norms.

13.1.2.2. Political and Bureaucratic Challenges

- **Political Resistance**: In some regions, the political environment may contribute to the resistance against divining. Governments and politicians may be reluctant to support methods that are not seen as **modern** or **scientifically validated**, especially in **urbanised** or **developed** areas where **modern engineering techniques** are heavily prioritised. This resistance can be compounded by **bureaucratic red tape**, where established procedures and channels must be followed, and divining is seen as an **extraneous or unnecessary step** in the process.

- **Sceptical Public Policy**: Public policy in many countries is based on the **principles of modernity** and **technological advancement**, which can exclude traditional practices like divining. National and regional water management strategies are often heavily influenced by **scientific institutions** and **international standards**, which focus on **geological surveying** and **hydrological mapping** rather than culturally embedded practices.

13.1.2.3. Strategies to Overcome Institutional Barriers

- **Policy Advocacy**: Engaging with policymakers to advocate for the **inclusion** of dowsing within water project planning could be an effective strategy. By presenting **evidence-based arguments** that show how dowsing improves water sourcing success rates and **reduces costs**, dowsing can be positioned as a **complementary technique**, not a replacement for modern engineering methods. Furthermore, **advocacy** can highlight how dowsing contributes to **community trust**, which is a key factor in the success of water projects.

- **Public-Private Partnerships**: Establishing **public-private partnerships** between local governments, **NGOs**, and **private-sector water management companies** could help bring divining into the fold. These partnerships can be a powerful way to implement more **inclusive** and **diverse** approaches to water sourcing, as the private sector is often more willing to take risks on innovative solutions. Moreover, **NGOs** working on water issues in rural areas may be able to support projects that integrate divining without the same level of scrutiny as government-funded projects.

- **Incorporating Divining into National Water Frameworks**: Governments and international bodies involved in water resource management should consider the **diverse contexts** in which water sourcing occurs. **Inclusive water resource management frameworks** could integrate **local knowledge systems** (like divining) into **national standards**, recognising them as complementary to more modern methods. Such frameworks could also offer **training programmes** that combine **traditional and modern techniques**, encouraging a more holistic approach to water resource planning.

13.1.3. Overcoming the Stigma Around Divining as Pseudoscience

Despite the **pragmatic successes** that have been documented in **divining-integrated projects**, the stigma around divining as **pseudoscience** remains a major hurdle to wider acceptance. This perception is driven by a combination of **cultural**, **historical**, and **scientific** factors, and overcoming it requires a multi-pronged approach.

13.1.3.1. Changing Perceptions

- **Cultural Shifts**: In many societies, divining is a **traditional practice** that is **deeply rooted** in local culture and spirituality. However, the **stigma of pseudoscience** often alienates younger generations and those exposed to more **scientifically-driven worldviews**. To counter this, it's important to **frame divining as a tool** that is **complementary** to modern science, rather than something that stands in opposition to it. This can be done by showcasing successful **community-driven projects** where divining was crucial in helping communities achieve access to water.

- **Public Education and Outreach**: Educating the public about the **effectiveness** and **history** of divining in practical applications is key to shifting perceptions. This can include **outreach programmes** that demonstrate how divining works in conjunction with modern engineering to **increase success rates**, reduce **environmental impacts**, and improve **community health outcomes**.

13.1.3.2. Documenting Success Stories

- **Empirical Data and Testimonials**: One of the most powerful tools for changing public perception is the **documentation of success stories**. If dowsing can be shown to result in **better outcomes** than traditional methods, this will gradually shift the view that dowsing is a mere superstition. The use of **quantitative data** (such as **higher success rates, lower costs, and increased water sustainability**) can help demonstrate that dowsing is not just folklore, but a **valuable resource** in certain contexts.

- **Collaborative Research Publications**: Collaborative publications between **scientists, engineers,** and **diviners** in reputable journals or at conferences can help **legitimise** divining as a **valid** and **effective technique** for water sourcing. By **bridging the gap** between scientific methodology and local knowledge, these efforts can show that divining is not just based on **intuition or guesswork** but can be a reliable and practical tool when used within a **multidisciplinary framework**.

13.2. Training and Capacity Building

One of the most significant challenges to the broader adoption of **divining techniques** in modern water management is the **lack of training** and **capacity building** among engineers, water professionals, and other stakeholders. In order for divining to be integrated effectively into water-sourcing projects, both **technical** and **cultural** knowledge must be shared between practitioners of traditional divining methods and those of modern engineering practices. This section explores how to **train engineers**, **water professionals**, and other key players to understand and incorporate traditional knowledge into modern water management strategies, as well as how to foster **cross-disciplinary collaboration** between diviners, engineers, and **NGOs**.

13.2.1. Training Engineers and Water Professionals

13.2.1.1. *Bridging the Gap Between Traditional and Modern Knowledge*

For engineers and water professionals to integrate **dowsing techniques** into their practice, they need both **technical training** and an understanding of how to **respect and utilise traditional knowledge** effectively. This training requires a **holistic approach** that values both scientific rigour and the lived experiences of local communities.

Key Training Elements

- **Introduction to Divining**: Engineers and water professionals need an **overview of divining**—what it is, how it works (from a traditional perspective), and how it has been used successfully in water projects. This could include understanding the various forms of divination, such as **dowsing rods, pendulum divining**, and **intuitive methods** employed by experienced diviners. Engineers should be introduced to **cultural practices** and **local knowledge systems** that have guided communities to water sources for centuries.

- **Collaborative Field Training**: Hands-on training is crucial. Engineers should spend time in the field, working alongside **experienced diviners** to understand their methods and perspectives. This could involve shadowing diviners during site assessments and observing how divining helps locate potential borehole sites. Such training will help engineers see the value of **local knowledge** in real-world contexts.

- **Combining Techniques in Planning**: Training should focus on how to **combine traditional divining methods** with **modern engineering tools** (such as **geophysical surveys**, **hydrogeological mapping**, and **remote sensing**). Engineers need to learn how divining can be used as an **initial screening tool**—to identify potential sites that can then be further investigated with more **scientific methods**.

- **Ethics and Sensitivity Training**: Engineers must also be trained in the **ethical considerations** of using divining techniques in their projects. They need to understand that **divining is a deeply cultural practice** and requires **respect** for local traditions. Training should emphasise how to approach diviners with **cultural sensitivity**, acknowledging their role in the community and ensuring that their contributions are valued and respected.

13.2.1.2. Creating Curriculum for Integrating Traditional and Modern Techniques

To institutionalise the integration of divining in engineering and water resource management, **formal training programmes** and **curricula** need to be developed. These could be offered through:

- **University-level Programmes**: Colleges and universities offering programmes in **environmental engineering, water**

resources management, or **geology** could create modules focused on integrating **traditional knowledge** into modern practices. This would not only educate future engineers about the value of traditional practices but also foster a broader acceptance of these techniques.

- **Professional Development Courses**: Organisations such as the **International Water Management Institute (IWMI)** or **NGOs** working in water resource management could offer specialised training for **current professionals**—engineers, hydrogeologists, and water management experts. These courses could include a combination of **theoretical content, case studies**, and **fieldwork** to ensure practical application.

- **Government Training Programmes**: Local governments could sponsor **training workshops** for engineers working in rural and underserved regions, where water scarcity is most acute. These workshops could be led by both **scientific experts** and **local diviners**, providing a balanced perspective and ensuring that solutions are culturally appropriate and scientifically sound.

13.2.2. Building a Cross-Disciplinary Understanding

The successful integration of **divining techniques** into water resource projects requires effective **collaboration** between **diviners, engineers,** and **NGOs**. Bridging the gap between these groups is essential for **mutual understanding** and creating a more **holistic approach** to water sourcing.

13.2.2.1. *Facilitating Collaboration Between Diviners and Engineers*

The relationship between **diviners** and **engineers** can be fraught with challenges due to differing worldviews and methodologies. Engineers may be sceptical of divining techniques, while diviners may not fully understand the engineering processes involved in **site testing** and **borehole drilling**. Effective collaboration will require **shared learning** and **mutual respect**.

Building Trust and Respect

- **Workshops and Dialogues**: To foster mutual understanding, regular **workshops** and **dialogue sessions** should be organised between diviners and engineers. These events can serve as platforms for both groups to share their knowledge and experiences. Engineers can learn about divining's **practical applications**, while diviners can gain insight into the **technical aspects** of drilling and water extraction.

- **Joint Field Visits**: Collaborative field visits are essential for **demonstrating how both methodologies** can complement one another. Engineers and diviners could visit sites together, where diviners suggest potential drilling locations, and engineers apply **scientific techniques** (such as **resistivity testing** or **geophysical surveys**) to verify the findings.

- **Mentorship**: Experienced diviners can mentor engineers and younger diviners alike, providing insight into how **traditional knowledge** can be applied in modern contexts. Engineers can also serve as mentors, teaching diviners about **modern engineering tools** and **water resource management techniques**, fostering cross-disciplinary understanding.

13.2.2.2. Role of NGOs in Bridging the Gap

Non-governmental organisations (NGOs), especially those focused on **water access** and **sustainability**, are well-positioned to act as intermediaries between **engineers, diviners,** and **local communities**. They play a crucial role in **facilitating training** and **fostering collaboration**.

NGO Functions

- **Mediators and Facilitators**: NGOs can mediate between **culturally distinct** groups, helping engineers understand the importance of **local knowledge** while encouraging diviners to see the value of **modern scientific methods**. They can organisefocus **groups**, **training sessions**, and **community workshops** that bring together engineers, diviners, and local stakeholders.

- **Project Coordination**: Many water access projects are spearheaded by NGOs, which can integrate **divining techniques** into their standard project design and implementation processes. NGOs can **design pilot projects** that combine both **engineering surveys** and **divining** to assess their combined effectiveness. These projects can then serve as examples of best practice.

- **Capacity Building for Local Communities**: NGOs can help train **local communities** to recognise the value of **integrating both methodologies** in water-sourcing projects. By emphasising the importance of **local knowledge** and the **scientific methods** used by engineers, NGOs can help communities develop a more **well-rounded approach** to water access.

13.2.2.3. Encouraging a Shared Vision for Water Management

Ultimately, the goal is to encourage all stakeholders to see the value of both **traditional** and **modern** techniques in addressing water scarcity. This requires a **shared vision** for **sustainable water management—** one that recognises that **science and tradition** are not mutually exclusive but can work together to meet the needs of communities.

Collaboration in Decision-Making

- **Inclusive Decision-Making**: Engineers, diviners, and local communities should all have a **voice** in **decision-making** processes related to water resource management. Creating **participatory forums** where different perspectives can be heard ensures that all stakeholders feel invested in the outcomes and are more likely to collaborate effectively.

- **Joint Problem-Solving**: When water issues arise, engineers and diviners should work together to **problem-solve**. For example, if an initial borehole fails to produce enough water, diviners can use their intuition to suggest alternative drilling locations, while engineers can test these sites with **scientific surveys**. This **collaborative approach** helps avoid the pitfalls of working in silos.

13.2.3. Strategies for Scaling Up Capacity Building

As the value of integrating **divining techniques** into water resource projects becomes more evident, efforts should be made to scale up **capacity building** and **training programmes** for engineers, water professionals, and diviners on a larger scale.

13.2.3.1. Institutionalising Cross-Disciplinary Training

- **Create Training Institutions**: Governments, NGOs, and universities can work together to establish **training centres** that provide cross-disciplinary education for both engineers and diviners. These institutions can offer certification programmes that certify professionals in **integrated water sourcing techniques**, which could be recognised by government agencies, development partners, and private sector firms.

13.2.3.2. Global Knowledge-Sharing Networks

- **International Collaboration**: There should be an emphasis on **global knowledge exchange**, where countries with experience in integrating divining with modern engineering can share their findings with others. Platforms like **UN-Water** or **the Global Water Partnership (GWP)** can help facilitate the exchange of **best practices** and **case studies**, enabling more widespread adoption.

- **Peer Networks and Conferences**: Engineers and diviners alike should have the opportunity to participate in **conferences** and **peer networks** focused on **innovative water solutions**. These events can provide a space for networking, sharing challenges, and learning about successful collaborations from around the world.

13.3. Regulatory and Policy Considerations

The **role of government** and **policy** in recognising and supporting the integration of **divining techniques** in water sourcing is critical to the wider adoption of this approach. Water management is typically governed by strict regulatory frameworks and national policies, often designed to ensure that water sources are safe, sustainable, and accessible to communities. Divining, which relies on traditional knowledge and practices, may face significant **regulatory and policy barriers** due to its perceived lack of scientific validation or formal recognition.

In this section, we explore the **challenges** and **opportunities** surrounding the integration of divining into formal water resource management policies. We examine how **governments** and **policymakers** can play a pivotal role in **recognising** and **supporting** divining practices as a **legitimate tool** for water sourcing, and how this can align with broader **water access** and **sustainability goals**.

13.3.1. Regulatory Challenges and Barriers

13.3.1.1. Lack of Formal Recognition for Divining Practices

In many countries, **national water policies** and **regulatory frameworks** are designed to prioritise scientifically validated and **technically proven** methods of water sourcing, such as **geological surveys**, **hydrogeological modelling**, and **drilling**. These methods are typically based on **modern scientific principles**, leaving little room for traditional practices like divining, which are not widely recognised by official institutions.

- **Absence of Legal Frameworks**: Dowsing is often not explicitly addressed in **water governance policies**. As a result, **local knowledge systems and**national water management frameworks, make it difficult for water professionals and communities to integrate them into formal water-sourcing strategies.

- **Regulatory Hurdles for Traditional Practices**: In many jurisdictions, **regulations** governing water drilling, borewell installations, and groundwater extraction are tightly controlled by government agencies, and only approved methods are accepted. Without **legal recognition** or **endorsement**, divining cannot be used to inform official water projects, even though it may be effective in certain local contexts.

- **Conflict with National Water Management Standards**: National water management plans and **international water protocols** (such as those promoted by the **World Bank** or the **United Nations Water** initiatives) often emphasise the use of **modern engineering technologies** and **scientific data** in water resource planning. Divining, by contrast, is seen as **subjective** and therefore fails to meet the **formalised standards** set by regulatory bodies.

13.2.3.2. Policy Gaps in Integrating Traditional Knowledge

There is also a **policy gap** in many regions where **traditional knowledge** and **modern scientific approaches** are seen as separate domains rather than complementary. While some countries have made strides in recognising the value of **indigenous knowledge** in conservation, agriculture, and climate resilience, **water management policies** often overlook the contribution of traditional water-sourcing methods like divining.

- **Policy Fragmentation**: Many water-related policies focus exclusively on **scientific data**, and in the absence of guidelines or incentives for integrating traditional practices, dowsing often remains excluded. For instance, policies regarding the **allocation of water rights**, **borehole drilling**, and **groundwater monitoring** do not typically take into account the use of **dowsing** as a **tool** for **preliminary site selection**.

- **Overemphasis on Quantitative Data**: Governments tend to prioritise **quantitative** and **technologically-driven data** when making decisions about water resource management. Divining, by nature, is often seen as a **qualitative** practice, which may result in it being dismissed or marginalised in official water planning processes.

13.3.1.3. Limited Policy Support for Interdisciplinary Approaches

In many regions, **water management** is governed by separate ministries or regulatory agencies focused on distinct aspects of the water cycle, such as **water supply, sanitation, environmental protection**, and **agriculture**. These **silos** often prevent an integrated, interdisciplinary approach to water management, which is necessary for recognising the value of both **scientific and traditional techniques**.

- **Lack of Coordination**: Without a framework that promotes cross-disciplinary collaboration, water resource planning often fails to consider how traditional methods like divining can complement modern engineering and scientific approaches. Divining is viewed in isolation, and the potential benefits of collaboration between **engineers**, **diviners**, and **local communities** are not recognised at the policy level.

13.3.2. Opportunities for Policy Reform and Government Support

13.3.2.1. *Creating Legal and Regulatory Pathways for Traditional Knowledge*

One of the most important steps in overcoming barriers to adopting divining techniques in water sourcing is the development of **legal frameworks** that formally recognise **local knowledge** and **traditional practices**. Governments can play a key role in fostering a more inclusive approach to water management by creating pathways for integrating divining into official water-sourcing processes.

Key Policy Recommendations

- **Incorporate Divining into National Water Policies**: Governments should consider **amending** existing water management policies to **recognise traditional practices**, including divining, as a **legitimate part** of the water-sourcing process. This could involve establishing **guidelines** for integrating **traditional knowledge** with **modern engineering techniques** in water projects.

- **Create Certification Processes for Diviners**: One way to bridge the gap between **traditional** and **modern practices** is to establish **certification programmes** for diviners, similar to those for professional engineers or hydrogeologists. This would allow diviners to receive **formal recognition** for their expertise and knowledge, and provide a **regulated pathway** for their involvement in water resource management.

- **Legal Protection for Traditional Knowledge**: Governments can also consider adopting **policies** that protect and support **indigenous knowledge systems** in water management, particularly in regions where **traditional knowledge** plays a vital role in **sustaining local water supplies**.

13.3.2.2. Promoting Policy and Research Collaboration

To break down silos and encourage interdisciplinary collaboration, governments can facilitate **joint research initiatives** between engineers, hydrogeologists, and local knowledge holders (including diviners). This would create a stronger case for the value of integrating divining with engineering methods and help develop a **comprehensive approach** to water resource planning.

Collaborative Research Platforms

- **Government-Funded Pilot Projects**: Governments could fund **pilot projects** that combine traditional divining with modern scientific techniques. These projects could test the effectiveness of **integrated approaches** and provide a body of evidence to support policy reform. Such initiatives could be part of broader **national water management plans** aimed at improving **water sustainability** and **climate resilience**.

- **Public-Private Partnerships**: In many cases, the private sector is at the forefront of **innovative water management solutions**. Governments can promote **public-private partnerships** to fund and implement integrated water projects that include **divining techniques** as part of the **preliminary site assessment** process.

- **International Collaboration**: Policymakers should also look to international examples and **global best practices** for integrating **traditional knowledge** into **modern water management**. International bodies such as the **United Nations, World Bank**, and **Global Water Partnership** can play a key role in supporting cross-border collaboration on projects that demonstrate the **success** of combining divining with **scientific water sourcing techniques**.

13.3.2.3. Building Capacity for Local Communities

Government policy should also recognise the importance of **community-led water management**. By **empowering local communities** to integrate their traditional knowledge with modern engineering, governments can foster **community ownership** of water resource projects, which leads to **more sustainable** and **long-term** water access solutions.

Policy Initiatives to Support Local Capacity

- **Community Water Programmes**: Governments can fund **community-driven** water projects that incorporate divining as part of the water-sourcing process. Such programmes would build **local capacity** and encourage **community participation** in the design, implementation, and maintenance of water infrastructure.

- **Incentivising Knowledge Sharing**: Policymakers should provide incentives for **knowledge sharing** between local communities and engineering professionals. By promoting **collaborative learning**, governments can foster a **more inclusive** and **holistic approach** to water resource management.

13.3.2.4. Enhancing Public Awareness and Support

In many cases, **public perception** plays a significant role in the success of water projects. Governments and policymakers need to **educate the public** on the benefits of **integrating traditional knowledge** with modern engineering practices, and how this can lead to more **sustainable** and **cost-effective water solutions**.

Public Awareness Campaigns

- **Advocacy for Integrated Water Solutions**: Governments can launch **public awareness campaigns** that highlight successful case studies where divining and engineering have worked together to improve water access. These campaigns can showcase the **scientific** and **cultural** value of integrating both approaches.

- **Community Engagement**: Policy initiatives should also promote **community engagement** in the water planning process, ensuring that the knowledge and experiences of local people are taken into account. This approach not only enhances the effectiveness of water sourcing but also builds **trust** and **collaboration** between communities and engineers.

13.4. Conclusions

13.4.1. Overcoming Barriers to Broader Adoption

To integrate dowsing more widely into **water resource management**, it is crucial to address the **scientific scepticism** and **institutional resistance** that currently exist. By **documenting successes, engaging with policymakers**, and fostering **collaborative research**, it is possible to show that dowsing, when used alongside modern engineering techniques, can improve **water sustainability**, reduce costs, and enhance **community engagement**.

Ultimately, overcoming the **stigma** around divining and positioning it as a **complementary tool** within the broader framework of water resource management will require a **cultural shift**—both in terms of how divining is perceived by the public and how it is integrated into **official systems** and **regulatory standards**.

13.4.2. Bridging Knowledge Systems

The key to overcoming the barriers to adopting divining in water resource management lies in **building strong, cross-disciplinary training programmes** and **fostering collaboration** between engineers, diviners, and local communities. By creating platforms for **mutual learning**, ensuring **cultural sensitivity**, and providing opportunities for **hands-on experience**, we can equip engineers with the skills they need to incorporate traditional knowledge into their work while respecting local cultures and traditions. This cross-pollination of knowledge is essential for **sustainable water solutions** that integrate the best of both **modern science** and **traditional wisdom**.

13.4.3. Supporting the Integration of Divining in Water Policy

Overcoming the regulatory and policy challenges surrounding the use of divining in water sourcing requires a combination of **legal recognition, interdisciplinary collaboration**, and **community engagement**. Governments can play a pivotal role in recognising **traditional knowledge** as a valuable tool in water management by creating **policy frameworks** that integrate both modern engineering and local practices. By **reforming policies, funding research**, and **promoting community-driven solutions**, governments can foster more **inclusive, sustainable**, and **effective water management practices** that benefit both **rural** and **urban communities**.

◆ ◆ ◆

Chapter 14

The Future of Water Resource Planning—Innovative Solutions for a Sustainable Future

14.1. Innovations on the Horizon:

- Exploring new technologies (e.g., AI, machine learning, and GIS) and how they might work in tandem with traditional methods like divining.

- The potential for combining ground-based divining with satellite technology for more precise mapping of groundwater.

14.2. Scaling the Model:

- How to replicate successful models in other regions and countries.

- The role of NGOs, governments, and the private sector in scaling up this integrative approach to water management.

14.3. The Role of Divining in Global Water Security:

- The long-term potential of divining is in helping to address the global water crisis, especially in regions with limited resources.

◆　◆　◆

14.1. Innovations on the Horizon

As the global demand for water continues to grow and climate change exacerbates water scarcity, innovative technologies and approaches are emerging to address these challenges. In particular, the combination of **modern technological advancements** and **traditional knowledge systems** has the potential to revolutionise**water resource planning**. In this chapter, we explore how cutting-edge technologies like **artificial intelligence (AI), machine learning (ML)**, and **geographic information systems (GIS)** can work in tandem with **traditional methods** like **divining** to create more **precise, efficient**, and **sustainable** water sourcing solutions. Additionally, we examine the promising potential of combining **ground-based divining** with **satellite technology** to enhance the **accuracy** of **groundwater mapping** and improve decision-making in water resource management.

14.1.1. Harnessing Modern Technologies in Water Resource Management

14.1.1.1. Artificial Intelligence (AI) and Machine Learning (ML)

AI and ML are transforming a wide range of industries, and **water resource management** is no exception. These technologies enable more **accurate predictions, better data integration**, and **smarter decision-making** in the management of water resources. In the context of integrating divining with modern engineering practices, AI and ML can be used to process large datasets, enhance predictive modelling, and improve the overall **accuracy** of water-sourcing efforts.

AI-Driven Predictive Modelling for Water Sources

- **Pattern Recognition**: By using **AI algorithms**, engineers and water professionals can analyse historical data from **hydrogeological surveys, groundwater levels**, and **climate trends** to identify **patterns** that indicate the presence of water.

AI models can process thousands of data points to predict where new water sources may be located.

- **Integrating Divining Data**: AI and ML could be used to incorporate **divining results** into predictive models. For example, if diviners identify potential water sources, AI algorithms could process these observations alongside geological and hydrological data to predict the likelihood of successful drilling and water yield at specific locations. AI could also identify potential correlations between **divining** and **subsurface water presence** that may not be immediately obvious through traditional engineering methods alone.

Machine Learning for Site Optimisation

- **Optimisation of Borewell Locations**: ML can help optimise the locations for drilling borewells by analysing both **divining** and **scientific data** (such as geophysical surveys or resistivity mapping). By combining these datasets, ML algorithms could **identify patterns** and **refine predictions** to minimise costly mistakes and increase the success rate of drilling projects.

- **Continuous Learning**: Machine learning systems can improve over time by **learning from past experiences**. As more data is collected from **divining techniques** and **drilling results**, the system becomes better at predicting where water is most likely to be found, based on **past successes** and **failures**.

14.1.1.2. Geographic Information Systems (GIS)

GIS technology is already widely used in environmental planning, urban development, and resource management. In water resource management, GIS provides powerful tools for mapping, spatial analysis, and decision support. By integrating **divining data** into GIS platforms, water professionals can create **dynamic maps** that combine

traditional knowledge with **geospatial data** for more accurate water resource planning.

GIS for Mapping Groundwater and Water Sources

- **Layered Maps**: GIS allows for the creation of **multi-layered maps** that can include various data sources, such as **hydrological data**, **soil composition**, and **divining results**. This enables engineers and diviners to see a **comprehensive view** of potential water sources, factoring in both **scientific surveys** and **local knowledge**.

- **Enhanced Site Selection**: When divining results are mapped onto GIS platforms, water professionals can assess whether diviners' recommendations align with **known geological features**, such as **fault lines** or **aquifers**. This can help identify which sites are most likely to yield successful results, making water-sourcing efforts more efficient.

Real-Time Data Integration

- **Remote Sensing and Ground-Based Data**: GIS can also integrate **satellite imagery** and **ground-based survey data** (such as **divining results** and **drilling test results**) to continuously update the mapping of groundwater resources. This real-time data integration allows water professionals to make more informed decisions as new information becomes available.

14.1.2. Combining Traditional Divining with Satellite Technology

One of the most exciting areas of innovation in water resource management is the combination of **ground-based divining** with **satellite technology** and **remote sensing**. This union promises to take the precision and effectiveness of **groundwater mapping** to a whole new level by combining **local knowledge** with **global monitoring tools**.

14.1.2.1. Satellite Technology for Groundwater Mapping

Satellites equipped with various sensing technologies are already providing valuable data for understanding groundwater systems, particularly in **arid** and **semi-arid** regions where access to groundwater is crucial. By using **remote sensing data**, satellites can measure **surface water levels**, **soil moisture content**, and even **subsurface features** that indicate the presence of groundwater.

Using Satellite Data for Hydrological Analysis

- **Gravitational Field Data**: Advanced **satellite missions**, such as the **GRACE** (Gravity Recovery and Climate Experiment) satellites, are capable of measuring changes in the Earth's gravitational field caused by changes in groundwater mass. This data helps monitor **groundwater depletion** and **accumulation**, making it a powerful tool for long-term water resource management.

- **Thermal Imaging and Soil Moisture: Satellite-based thermal imaging** can be used to detect **moisture levels** in the soil, which may suggest the presence of subsurface water. When combined with divining methods, satellite data could provide **concurrent validation** of groundwater availability in areas where traditional methods are often used in isolation.

14.1.2.2. How Divining and Satellite Technology Could Work Together

While satellite data offers broad-scale coverage, **divining** provides localised, **on-the-ground insights** that may be difficult to capture from space. Combining these two approaches can create a more **precise, data-driven**, and **culturally sensitive** model for groundwater exploration.

Integrating Ground-Based Dowsing with Satellite Data

- **Preliminary Site Identification**: Diviners can use their skills to identify **potential groundwater locations** based on local knowledge, and these areas can be further investigated using satellite and remote sensing technologies. For example, divining could suggest areas where water is likely to be found, and satellite data can help **validate** or **refine** these locations by providing **geospatial information** about soil moisture or vegetation patterns that correlate with groundwater presence.

- **Combining Data Sets**: Engineers and water professionals could integrate **divining data** (such as diviners' observations and recommendations) with **satellite imagery, GIS maps**, and **hydrological modelling** to create a **comprehensive map** of groundwater resources. This combined data approach would enable water projects to be more **precise**, reducing the likelihood of failed borewell drilling and increasing overall success rates.

- **Remote Monitoring and Data Validation**: Once a potential water source is identified through divining and satellite data, **remote monitoring tools** (such as **drones, satellite-based groundwater sensors**, and **IoT** devices) could be used to continuously track water availability and **groundwater changes**. This real-time data would validate the success of previous efforts and help engineers adjust drilling strategies accordingly.

14.1.3. The Future of Collaborative Water Resource Management

As technology continues to evolve, the future of **water resource planning** will likely involve a **collaborative** approach that combines **modern scientific techniques** with **traditional knowledge**. This cross-disciplinary model could help create more **sustainable, resilient, and inclusive** water systems that benefit local communities while addressing global water challenges.

14.1.3.1. Developing a Holistic, Integrated Approach

- **Co-Designing Solutions**: The integration of **AI, ML, GIS**, and **divining** will require close collaboration between **scientists, engineers, diviners**, and **local communities**. By co-designing water resource management solutions, all stakeholders can contribute their knowledge and expertise, creating more **contextually appropriate, effective**, and **sustainable** water-sourcing strategies.

14.1.3.2. Capacity Building and Training

- As these technologies become more integrated into water resource planning, it will be essential to provide **training programmes** that equip water professionals, diviners, and local communities with the skills to work together. This could involve a combination of **technical training** on AI, ML, GIS, and satellite technologies, alongside **cultural training** on the value of traditional knowledge and the importance of community engagement.

14.1.3.3. Policy Support for Innovation

- Governments and policymakers will need to develop **policies** that **support** and **facilitate** the use of **innovative technologies** in combination with **traditional practices**. This includes **legal recognition** of divining as a complementary tool, as well as creating **funding mechanisms** to support **research, pilot projects**, and **capacity-building initiatives**.

14.2. Scaling the Model

The integration of **traditional knowledge** (such as divining) with **modern engineering and technological** tools represents a promising approach to water resource management. However, for this model to have a lasting, global impact, it must be **scaled** across different regions, countries, and communities. Scaling up requires overcoming logistical, financial, cultural, and institutional challenges, but it also offers significant opportunities for **sustainable water management** at a larger scale.

In this section, we explore how successful models of integrated water management can be **replicated** across various regions and countries, and the critical roles that **NGOs, governments**, and the **private sector** play in **scaling** these efforts.

14.2.1. Replicating Successful Models in Other Regions

14.2.1.1. Identifying Transferable Elements of the Model

The key to **scaling** an integrated approach to water resource management is identifying the **core components** of the model that can be successfully applied in different geographical and cultural contexts. Successful water projects that combine **divining** with **engineering** and **technological tools** often share several common elements that can be adapted to other regions.

Key Elements for Replication

- **Collaboration Across Disciplines**: One of the core principles of successful integrated water management projects is collaboration between **local communities, engineers,**

hydrologists, diviners, and **government bodies**. This collaborative approach should be prioritised in scaling efforts.

- **Local Knowledge Integration**: The inclusion of local knowledge (such as divining) should be seen not just as an add-on, but as an essential component of the water-sourcing process. Successful projects integrate the **wisdom** of local people into the **scientific process**, ensuring that culturally relevant and context-specific knowledge is acknowledged and used.

- **Technological Support**: The combination of **AI, GIS, remote sensing**, and **groundwater modelling** has been shown to improve the accuracy of water sourcing in regions with scarce or unpredictable groundwater resources. The adaptability and scalability of these technologies make them ideal for replication.

- **Flexible and Adaptive Implementation**: Different regions face different challenges, so the model must be **adapted** to local conditions. In scaling, it is critical to **customise** the approach to the **specific needs** and **resources** of each region, adjusting for terrain, climate, water scarcity levels, and community dynamics.

14.2.1.2. Lessons Learned from Successful Pilot Projects

To effectively replicate successful integrated models, lessons learned from **pilot projects** are invaluable. These pilots typically involve smaller-scale interventions that can be refined and tested before expanding to larger areas. Some examples of successful models for replication include:

- **Rural India**: In rural India, projects that combine **divining** with **hydrogeological surveys** and **remote sensing** have been effective in locating water sources where conventional methods often fail. These projects rely on **local diviners**, working alongside **scientists** and **engineers**, to pinpoint water-bearing locations in difficult-to-reach rural areas.

- **Sub-Saharan Africa**: In some African countries, local diviners have worked with **NGOs** and **engineering teams** to locate water sources for **boreholes** in drought-prone areas. Satellite data and **GIS** technology are then used to verify divining predictions and **optimise drilling locations**. The integration of traditional practices with **modern technology** has increased the success rate of drilling and decreased the cost of water sourcing.

- **Latin America**: Several community-led water management initiatives in parts of Latin America have successfully incorporated **local knowledge** in water sourcing. By involving both **indigenous knowledge** and **modern engineering**, these models have been able to locate and manage water resources sustainably in regions where freshwater is scarce.

These case studies show that with careful planning and **contextual adaptation**, the model of integrating divining with modern technologies can be replicated in different parts of the world.

14.2.2. The Role of NGOs, Governments and the Private Sector

Scaling the integrative approach to water management requires the active involvement and collaboration of **multiple stakeholders**, including **NGOs, governments**, and the **private sector**. Each of these actors plays a distinct and complementary role in facilitating the broader adoption and successful implementation of these innovative solutions.

14.2.2.1. NGOs and Community-Based Organisations (CBOs)

Non-governmental organisations (NGOs) and community-based organisations (CBOs) are often the **catalysts** for water resource initiatives, especially in **rural, remote,** or **developing** regions. Their role in scaling the model is critical, as they provide the essential **community links, on-the-ground knowledge,** and **networking capacity** needed for successful implementation.

NGO Contributions

- **Capacity Building**: NGOs often play a significant role in training **local communities, diviners,** and **water professionals** on how to integrate traditional knowledge and modern technologies. By building the **capacity** of local stakeholders, they help ensure that solutions are sustainable in the long term.

- **Facilitating Partnerships**: NGOs can act as mediators between local communities, governments, and the private sector, helping to **coordinate efforts** and align goals across different sectors. They also play a critical role in ensuring that **local knowledge** is respected and integrated into planning processes.

- **Monitoring and Evaluation**: NGOs can monitor and evaluate the impact of water projects, ensuring that scaling efforts are **successful** and **sustainable**. They help identify best practices and refine approaches to address unforeseen challenges.

14.2.2.2. Governments and Policymakers

Governments play an essential role in the **policy, regulatory,** and **funding** aspects of scaling up integrated water management models. **National and local governments** are typically responsible for setting the legal and regulatory frameworks that govern **water access**, **groundwater management**, and **infrastructure development**.

Government Roles in Scaling

- **Policy and Regulatory Support**: Governments must create **policies** that encourage the integration of **local knowledge** into water resource management, recognise**divining** as a **legitimate tool**, and incentivise the use of **modern technologies** in combination with traditional methods. **Legal frameworks** should also be developed to ensure that water rights, drilling permits, and borewell installations are regulated in a way that supports these integrative practices.

- **Funding and Investment**: Scaling requires substantial financial resources. Governments are often in the position to provide **funding**, **subsidies**, or **grants** to support the development and replication of successful models. This could include **investing in research** and **technology** to improve **borehole success rates**, as well as **supporting capacity-building initiatives** for local communities.

- **Public Awareness and Advocacy**: Governments can help raise awareness of the importance of integrating **traditional knowledge** with **modern technologies** by engaging in **public education campaigns**. This can help reduce scepticism and build **trust** in the methods among local populations.

14.2.2.3. Private Sector and Technological Innovation

The **private sector**, especially companies involved in **technology**, **water engineering**, and **infrastructure development**, plays a crucial role in scaling up integrative water management practices. The private sector can provide the necessary **technology**, **expertise**, and **investment** to improve the accuracy, cost-effectiveness, and sustainability of these projects.

Private Sector Contributions

- **Technology and Infrastructure**: The private sector is often at the forefront of **innovative technologies**, such as **satellite imaging, remote sensing, AI-powered predictive models**, and **groundwater sensors**. These technologies can significantly improve the precision of **water sourcing**, reduce costs, and enhance the overall success of **integrated water management** models.

- **Public-Private Partnerships (PPPs)**: Successful scaling often requires the formation of **public-private partnerships** (PPPs), where the expertise and resources of the private sector are combined with the regulatory and social responsibility mandates of the public sector. These partnerships can help overcome financial and logistical barriers, especially in low-income regions.

- **Investment in Sustainable Solutions**: Private companies can also contribute by investing in **sustainable water technologies**, offering **affordable solutions**, and exploring **business models** that support **long-term water resource management**. Companies may also be incentivised by the **social impact** of their investments in underserved communities.

14.2.3. Collaborative Scaling Approach: A Global Network for Water Management

To successfully scale integrated water management practices globally, it is crucial to establish a **global network** of **collaborative partnerships** between **NGOs, governments, private sector actors, scientists**, and **local communities**. This network can serve as a platform for **sharing best practices, coordinating efforts**, and **mobilising resources** to scale innovative water management solutions.

Key Steps for Collaborative Scaling

- **International Knowledge Exchange**: Hosting **workshops**, **conferences**, and **knowledge-sharing platforms** where stakeholders can exchange lessons learned, successful case studies, and innovative strategies will help foster cross-border collaboration.

- **Global Advocacy and Policy Influence**: Organisations like the **United Nations** or the **World Bank** can help advocate for the adoption of integrative water management models in global policy frameworks, encouraging national governments to prioritise**cross-disciplinary solutions**.

- **Funding Mechanisms**: International and national funding bodies (e.g., the **Global Green Fund** or the **World Water Fund**) can provide **financial support** to scale successful projects, ensuring that resources are accessible for implementation in both urban and rural areas.

14.3. The Role of Divining in Global Water Security

As the world faces a growing water crisis, exacerbated by **climate change**, **population growth**, and **increasing urbanisation**, innovative solutions to water sourcing and management are becoming more critical than ever. One such solution, which has often been dismissed or overlooked by mainstream scientific communities, is **divining** (or **dowsing**)—the practice of locating water sources using traditional, intuitive methods. While the practice has historically been seen as folklore or pseudoscience, growing interest in integrating **traditional knowledge** with **modern engineering** is highlighting divining's potential as part of a broader strategy for addressing **global water security**.

In this section, we explore the **long-term potential** of **divining** as a tool to help tackle the **global water crisis**, particularly in **regions**

with limited resources, and examine how it can complement scientific techniques to create more **sustainable**, **cost-effective**, and **culturally appropriate** water sourcing solutions.

14.3.1. Divining as a Tool for Sustainable Water Sourcing in Resource-Limited Regions

14.3.1.1. Addressing Water Scarcity in Developing Regions

Water scarcity is one of the most pressing challenges in many **developing regions**, particularly in **sub-Saharan Africa**, parts of **South Asia**, and **Latin America**. In these areas, access to **clean water** is limited, and conventional methods for finding groundwater sources can be expensive, technically difficult, and resource-intensive. Traditional engineering methods, such as **hydrogeological surveys** and **borehole drilling**, often rely on advanced equipment, skilled professionals, and substantial capital investments, all of which are difficult to mobilise in regions with limited financial resources.

Divining offers a **low-cost, locally accessible** alternative. Skilled diviners, who typically have a deep understanding of the local environment and water systems, can pinpoint potential groundwater sources without the need for expensive equipment. This can be especially useful in areas where:

- **Geological data is scarce** or difficult to interpret.

- **Remote regions** lack access to advanced technology or expertise.

- **Financial constraints** make large-scale scientific exploration and drilling projects unfeasible.

In such contexts, divining can serve as a **complementary tool**, helping to guide **drilling efforts** and **optimise resource allocation**, potentially saving both **time** and **money**. When used in tandem with modern

geophysical surveys, **GIS mapping**, and **satellite data**, divining can improve the **accuracy** of water sourcing while keeping costs low.

14.3.1.2. Low-Cost and Low-Tech Solutions for Rural Communities

In many rural or isolated communities, access to **advanced technology** and **scientific expertise** is limited. Yet these communities often have deep historical and cultural knowledge of the landscape and water sources. Divining, being an **inexpensive** and **simple** practice, is often a **more accessible** solution in such contexts compared to sophisticated technologies.

By integrating dowsing into local water resource management strategies, communities can:

- **Increase local participation** and **ownership** in water projects.

- **Empower local diviners** with the knowledge and tools to guide water-sourcing efforts.

- Use **traditional methods** alongside more **modern engineering** to create a **sustainable and locally driven solution**.

Moreover, diviners often have a keen intuition about the **local terrain** and **environmental factors**, which can be particularly useful in **dryland** or **semi-arid regions**, where the ability to read subtle signs of water presence can make a difference in successfully locating water sources.

14.3.2. Enhancing Water Sustainability Through Integrative Practices

14.3.2.1. Complementing Scientific Methods with Traditional Knowledge

While divining is not widely accepted by mainstream science, there is growing recognition that combining **traditional knowledge** with

modern technologies can offer a more **holistic, inclusive**, and **sustainable** approach to water management. In regions where water resources are limited, such as **desert zones, mountainous regions**, and **coastal areas**, divining can be a valuable tool in **augmenting** the work of hydrogeologists, engineers, and environmental scientists.

When divining is integrated into **modern water resource planning**, the outcomes can be **improved** in several ways:

- **More accurate site selection** for borewells, reducing the risk of costly drilling failures.

- **Better resource allocation**, by focusing drilling efforts on sites identified by divining in conjunction with scientific data.

- **Lower operational costs**, as divining can reduce the need for expensive, large-scale geological surveys or prolonged drilling efforts.

14.3.2.2. A Cost-Effective Approach for Water Security

In resource-limited settings, the cost-effectiveness of water projects is crucial. Many countries in the Global South struggle with **limited financial resources** and must find ways to maximise the impact of their water infrastructure investments. Divining offers a **cost-effective, rapid**, and **low-tech** solution to **locating water**, making it an important tool for **scaling up water resource access**.

In addition, divining can help prioritise water projects that have the **highest chance of success**. By reducing the risk of dry boreholes, divining can help direct efforts towards sites with the greatest likelihood of yielding fresh water, saving communities both **time** and **money**. This is particularly important in **remote** or **rural areas** where the failure of a borehole can have significant consequences on both **community health** and **economic stability**.

14.3.3. Long-Term Potential of Divining in Global Water Security

14.3.3.1. *Supporting Resilience to Climate Change*

As the global climate crisis accelerates, many regions are facing **more erratic weather patterns**, including **droughts, floods**, and **shifting rainfall patterns**. This poses a significant threat to water security, especially in already **water-stressed regions**. Divining, with its **localised knowledge** of the landscape and water systems, can help improve **climate resilience** by:

- **Identifying hidden water sources** that might otherwise go unnoticed by traditional surveying methods.

- Helping communities **adapt to changing climate patterns**, especially in areas where conventional methods of locating groundwater are less effective.

- Complementing **climate adaptation strategies** that prioritise sustainable and long-term water access.

As weather patterns become less predictable and water resources become scarcer, **integrating traditional methods like divining** can help communities better **adapt** to these changes by providing an additional layer of insights into water availability and strengthening local **resilience.**

14.3.3.2. *Providing Global Solutions for Local Water Needs*

Divining may never replace modern hydrogeological methods, but it offers an important complement, especially in places where traditional engineering is difficult to implement or cost-prohibitive. The **global water crisis** is not a one-size-fits-all problem, and **local solutions** are often the most effective. By integrating **local knowledge**, such as divining, into **global water management strategies**, a more **diverse, inclusive**, and **innovative** set of solutions can emerge.

Divining can also help bridge the gap between **scientific approaches** and **local communities**, creating a more **collaborative, trust-based** approach to water resource management. This is especially important in areas where **local populations** have historically been excluded from decision-making processes or where there is mistrust between **scientists, governments**, and **local communities**.

14..3.4. Building Trust and Inclusivity for Broader Acceptance

14.3.4.1. Overcoming Scepticism in the Scientific Community

One of the challenges in scaling divining for global water security is overcoming the scepticism that many in the **scientific community** have towards traditional methods. However, as more **case studies** and **evidence** emerge, particularly from regions where divining has been successfully integrated with scientific practices, it is becoming increasingly clear that **divining** can play a role in **improving water sourcing** and **enhancing sustainability**.

14.3.4.2. Promoting Cross-Disciplinary Collaboration

For divining to be accepted and integrated into mainstream water management practices, it is crucial to promote **cross-disciplinary collaboration** between **scientists, engineers, local communities**, and **diviners**. This requires fostering **mutual respect, knowledge sharing**, and a willingness to recognise that **local wisdom** can contribute meaningfully to **scientific research** and **policy development**. Establishing frameworks that **legally recognise** divining and encourage collaboration between diverse stakeholders will be critical to **scaling up** its impact.

14.4. Conclusions

14.4.1. The Future of Synergy and Sustainability

The future of water resource planning will increasingly rely on the **synergy** between modern technologies like **AI, GIS**, and **satellite imaging** with **traditional knowledge systems** like **divining**. By combining the **local insights** of diviners with the **global reach** and **precision** of modern technology, we can create more effective and sustainable solutions for water sourcing. This integrated approach promises not only to **improve efficiency** and **reduce costs** but also to build **trust** with local communities and ensure that water resources are managed in a way that is both **scientifically sound** and **culturally sensitive**. The future of water resource planning is one of **collaboration, innovation**, and **sustainability**, where **science** and **tradition** work together for the common good.

14.4.2. A Roadmap for a Sustainable Water Future

Scaling integrated water resource management models that combine **traditional practices** with **modern technologies** offers a powerful pathway to address global water challenges. By leveraging the strengths of **NGOs, governments**, and the **private sector**, and by **adapting successful models** to diverse regional contexts, we can foster a more **equitable, sustainable**, and **resilient** future for water management across the globe. Through **collaboration, innovation**, and **inclusive approaches**, this integrative model can contribute to solving one of the world's most pressing challenges: ensuring access to clean and sustainable water for all.

14.4.3. Divining's Role in a Sustainable Water Future

The long-term potential of divining as a tool for water security is vast, particularly in regions where access to modern technology and resources is limited. While it may not be the sole solution, **divining** can complement **modern engineering techniques** and **scientific tools**, offering a more **cost-effective, localised**, and **culturally sensitive** approach to locating groundwater. By integrating divining into broader water management strategies, we can help ensure **sustainable water access** in **resource-limited** and **climate-vulnerable regions**, improving the resilience and security of communities worldwide.

As we look to the future, the key will be to embrace**innovative solutions** that combine the **best of both worlds**—the **wisdom of tradition** with the **advances of modern science**—to meet the urgent water needs of **communities across the globe**.

◆　◆　◆

Summary: A New Paradigm in Water Resource Planning

(A). Recap of Key Insights:

- How integrating traditional and modern practices can offer reliable, sustainable water solutions.

(B). Call to Action:

- Encouraging engineers, NGOs, and policymakers to adopt an open-minded, interdisciplinary approach to water resource planning.

(C). Vision for the Future:

- A hopeful outlook on how combining different approaches can lead to more equitable and resilient water systems, especially in rural communities.

◆ ◆ ◆

(A). Recap of Key Insights

As the world grapples with escalating water crises, particularly in developing regions and those facing the impacts of climate change, the search for sustainable, reliable, and cost-effective water-sourcing solutions has never been more urgent. In this context, the integration of **traditional knowledge**—such as **divining**—with **modern engineering practices** and **advanced technologies** offers a promising, **innovative pathway** for addressing the complex challenges of **water scarcity** and **resource management**.

Here are the **key insights** that have emerged from our exploration of this integrated approach:

A.1. Bridging the Gap Between Tradition and Technology

For centuries, **dowsing** has been used by local communities to locate water sources based on **intuition** and **cultural knowledge** of the land. Despite scepticism from the scientific community, the growing interest in integrating **traditional practices** with **modern technologies** underscores the potential for these methods to complement each other in water management.

Divining is not a replacement for **hydrogeological surveys**, **remote sensing**, or **GIS mapping**, but it can enhance their effectiveness. When divining is used alongside **modern engineering techniques**, the result is a more **accurate**, **cost-effective**, and **community-centred** approach to locating and managing water resources. This **synergy** offers a more comprehensive understanding of the **local environment** while addressing the **practical constraints** of many regions, especially those with limited access to advanced technologies.

A.2. Cost-Effectiveness and Accessibility

In regions with **limited resources**, where **scientific equipment** and **expertise** are often unavailable, divining offers a **low-cost** and

locally accessible solution for finding groundwater sources. It helps **optimise drilling efforts, reduce the risk of failure**, and **minimise the financial burden** of water-sourcing projects. This makes divining an invaluable tool, especially in **remote** and **rural** areas where access to modern water technologies is often restricted.

Divining also plays an important role in ensuring **local participation** and **empowerment** in water resource projects. Involving local diviners who have intimate knowledge of the terrain and water systems builds **trust** and encourages **community ownership**, which is critical for the long-term sustainability of water projects.

A.3. Enhancing Sustainability and Resilience

The integration of traditional methods with modern technologies enhances the **resilience** of water-sourcing projects, particularly in the face of **climate change** and **variable weather patterns**. Divining can help **identify hidden water sources** and provide insights into groundwater behaviour in ways that modern technologies might miss. This is especially important in regions where **climate-induced water stress** is increasing, and where water resources are becoming **more unpredictable** and **difficult to locate**.

By combining the **predictive power** of technologies like **satellite imagery** and **AI-driven models** with the **local intuition** of diviners, communities can develop more **adaptive** and **sustainable water management strategies**. This holistic approach enables water resource planners to make more **informed decisions** that are **scientifically sound** and **culturally sensitive**.

A.4. A Collaborative, Cross-Disciplinary Future

One of the most significant takeaways from this exploration is the importance of **collaboration** across **disciplines**. To fully unlock the potential of integrating divining with modern water management practices, engineers, scientists, diviners, local communities, NGOs, and governments must work together in **partnership**. By fostering a **shared understanding** and **mutual respect**, these diverse stakeholders can collaborate to build more effective, adaptable, and context-specific water solutions.

This **cross-disciplinary** approach helps to break down **institutional silos** and encourages a more **inclusive** form of water resource management, where all forms of knowledge—whether **scientific** or **traditional**—are valued and utilised.

A.5. A Path Towards Global Water Security

The future of water resource planning lies in creating **innovative**, **scalable**, and **sustainable solutions** that address the **complex realities** of water scarcity and **equitable access**. Integrating divining into water resource planning, particularly in **resource-constrained regions**, offers a way forward that balances **modern technology** with **local wisdom**, creating a system that is **both scientifically rigorous** and **culturally appropriate**.

In the long term, this integrated approach could play a significant role in **mitigating the global water crisis**, especially in **water-stressed** and **climate-vulnerable regions**. By bringing together the **best of both worlds**—traditional knowledge and modern science—we can develop a more **holistic**, **resilient**, and **sustainable** model for managing water resources and improving **global water security**.

(B). Call to Action: Embracing an Open-Minded, Interdisciplinary Approach to Water Resource Planning

As we stand at the crossroads of a global water crisis, with **climate change**, **population growth**, and **resource depletion** all threatening the future of freshwater availability, the need for **innovative, adaptable**, and **inclusive solutions** has never been more urgent. The challenge we face is not just about finding new ways to access water, but about rethinking how we approach water resource planning itself.

In the pursuit of sustainable water management, it's essential that we look beyond the confines of conventional methods and embrace an **interdisciplinary approach** that **integrates diverse knowledge systems**—including **traditional knowledge**—with **modern engineering, science,** and **technology. Divining,** a practice long regarded with scepticism by the scientific community, offers an excellent example of how **cultural wisdom** and **innovative engineering** can work together to solve complex water challenges.

B.1. Encouraging Open-Mindedness in Engineers, NGOs, and Policymakers

We urge **engineers, water professionals, NGOs,** and **policymakers** to adopt an **open-minded** approach to water resource planning—one that welcomes **cross-disciplinary collaboration, local participation,** and **inclusive decision-making.** By fostering **respectful partnerships** between **scientists, diviners, local communities,** and **governments,** we can create **holistic** water solutions that are both **scientifically grounded** and **culturally relevant.**

The water crisis is not only a technical challenge but also a **social, cultural,** and **economic** one. In regions where **advanced technologies** are scarce and **financial resources** are limited, **traditional practices** like divining offer valuable insights into local hydrology and groundwater potential. However, divining alone is not enough. Its true potential is realised when it is **integrated** with modern **hydrological data, geological surveys,** and **remote sensing technologies.** This combination will provide the **best of both worlds**—empowering local communities with both **scientific expertise** and **cultural knowledge.**

B.2. A Collaborative, Inclusive Water Future

It is essential for **engineers** and **water professionals** to move beyond a **narrow view** of water management and explore **alternative perspectives**. By creating **collaborative spaces** for **local diviners, scientists**, and **NGOs** to exchange knowledge and learn from one another, we can **bridge the gap** between **traditional knowledge** and **modern technology. Community engagement**, trust-building, and **mutual respect** are key to creating **sustainable, scalable**, and **culturally appropriate** water solutions that will stand the test of time.

B.3. Policy and Institutional Support

For this new paradigm in water resource planning to take root, **policymakers** must play a pivotal role in supporting and institutionalising interdisciplinary approaches. This means **recognising the value of traditional knowledge** and **integrating it into national and regional water policies**. Governments should consider creating **legal frameworks** that acknowledge the contributions of local **diviners**, as well as other traditional water management practices, alongside modern scientific methodologies.

B.4. The Time to Act is Now

The path forward requires a **shift in mindset**—a recognition that **water security** cannot be achieved through one-size-fits-all solutions, but through a diversity of tools, practices, and **knowledge systems. Innovation, collaboration**, and **adaptability** will be the keys to solving the **global water crisis**.

We call on **engineers, NGOs, policymakers**, and **local communities** to embrace a **shared responsibility** in the sustainable management of water resources. It is time to move beyond silos and open the door to a **new paradigm**—one that blends the **wisdom of**

tradition with the **advances of modern science**, to **ensure access to clean water** for all, especially for those who need it most.

(C). Vision for the Future

As the world faces mounting challenges around water scarcity, climate change, and growing populations, the need for more **equitable, resilient**, and **sustainable** water systems has never been more urgent. The **traditional approach** to water resource management—largely driven by **engineering solutions, hydrological surveys**, and **modern technologies**—has undoubtedly made significant strides. Yet, it often overlooks the deep **local knowledge, cultural practices**, and **community-based solutions** that have existed for centuries.

Looking towards the future, we can envision a **new paradigm** for water resource planning—one where **traditional knowledge**, such as **divining**, is **complemented** by **modern scientific techniques**. This vision is not merely about the **integration of technologies** but about a **shift in mindset**—from **top-down, one-size-fits-all solutions** to a more **collaborative, inclusive, and localised approach**. It's about recognising that **diverse knowledge systems** can work together to create **water systems that are not only effective** but also **equitable** and **resilient**—especially in **rural communities** where the need for clean, reliable water is often most pressing.

C.1. Empowering Rural Communities with Local Knowledge

The future of water resource planning must prioritise the **empowerment of rural communities**. Many of these communities already possess a deep understanding of their local environment, including natural water sources, seasonal patterns, and how to manage limited resources. By embracing traditional methods like **divining**, we acknowledge and value the wisdom of these communities. Rather than dismissing local practices as **outdated** or **unscientific**, we recognise them as valuable **tools** in the broader toolkit for water management.

When **diviners** work alongside engineers and hydrogeologists, they can help identify **hidden water sources** and provide crucial insights into the behaviour of groundwater that may not be captured by traditional scientific methods alone. In turn, modern technology—such as **remote sensing, satellite imagery**, and **hydrogeological surveys**—can provide a broader, **data-driven understanding** of water systems that can be applied more effectively.

This **cross-pollination of ideas** will not only help **enhance the efficiency** of water-sourcing efforts but also foster a **stronger sense of ownership** and **local pride** in water management projects, leading to more **sustainable, long-term solutions**.

C.2. Building Resilient and Adaptable Water Systems

As we look to the future, it is clear that **resilience** is the key to overcoming the challenges of **climate change, water scarcity**, and **environmental stress**. The future of water systems must be **adaptable** and **flexible**, able to respond to shifting weather patterns, changing populations, and unpredictable resource availability.

Combining **divining** with **modern technologies** can help build **resilient water systems** that are better equipped to adapt to these uncertainties. Traditional knowledge, with its focus on **local conditions, adaptive techniques**, and **historical understanding** of water resources, can provide the **foundation for more flexible, community-driven solutions**. In turn, advanced technologies can **scale** these solutions, providing the **data and insights** needed to make more informed decisions at the regional and national levels.

For example, **integrating divining results with remote sensing data** could help communities pinpoint underground water reserves that might not be visible to traditional surveying methods. In times of **drought** or **water stress**, this approach can ensure that **water-sourcing projects** are more **efficient, cost-effective**, and **sustainable** over the long term.

C.3. Ensuring Equitable Access to Water for All

A crucial aspect of this new paradigm is ensuring that water solutions are **equitable**, particularly in **rural** and **marginalised communities**. Water is a **fundamental human right**, and its availability should not be determined by wealth, status, or geography. Yet, in many parts of the world, **rural communities**—especially those in **remote** or **dryland regions**—often find themselves without reliable access to clean water.

By **integrating traditional knowledge** with modern water management practices, we can ensure that even the most **resource-limited communities** have access to clean, safe, and reliable water. Traditional methods like divining offer a **low-cost, accessible** approach that can complement expensive, large-scale infrastructure projects. This combination can help bridge the gap between communities with limited resources and the **technological advancements** that are more readily available in urban areas.

This approach encourages **local ownership** of water projects, with communities themselves actively participating in the **decision-making process**. When local people have a stake in the management of their own water resources, they are more likely to adopt **sustainable** practices and invest in **long-term solutions**.

C.4. Fostering Collaboration for a Global Water Future

Looking ahead, one of the most exciting aspects of integrating traditional and modern water management practices is the potential for **global collaboration**. The **global water crisis** is one of the most pressing challenges humanity faces, and no single country or organisation can solve it alone. To make a meaningful impact, there must be a shift towards more **collaborative, interdisciplinary** approaches to water resource planning.

By fostering **dialogue** between **engineers, hydrologists, diviners, NGOs, governments**, and **local communities**, we can create a **global**

water network that combines the best of both **scientific rigour** and **local knowledge**. This network will be more **flexible, innovative,** and **inclusive**, ensuring that the needs of the most vulnerable communities are met while also promoting **global water security.**

International institutions and **policymakers** must take an active role in encouraging these **cross-disciplinary partnerships.** By recognising the value of **traditional knowledge** in water resource planning and supporting **inclusive policies** that embrace both **modern** and **traditional solutions**, we can work towards a **water-secure future** for all.

C.5. A Hopeful Path Forward

The future of water resource planning lies in the **fusion of traditional and modern knowledge**—a **new paradigm** that is **equitable, resilient,** and **inclusive.** This vision is not just about **technology** or **cultural practices** in isolation but about how **both** can come together to create more **sustainable, adaptable,** and **community-driven** water systems.

In **rural communities**, where access to water is often the most critical, the integration of **local knowledge** and **scientific expertise** can lead to more **effective** and **lasting solutions.** By recognising that **traditional practices** like divining offer valuable insights into local water systems, we can begin to forge a path towards **greater equity** in water access—ensuring that no community is left behind.

This future requires **collaboration, openness,** and a willingness to challenge traditional boundaries in water resource management. Let us embrace this opportunity, work together, and **build a water-secure world** for future generations.

◆　◆　◆

(D). Conclusion

D.1. A New Paradigm in Water Resource Planning

The challenges of the **global water crisis** require a new approach—one that transcends the traditional boundaries of science, engineering, and local knowledge. By recognising the **value of traditional practices** like divining and integrating them with **modern scientific techniques**, we are not only improving the effectiveness and efficiency of water sourcing but also fostering **collaboration, community involvement,** and **sustainable practices**.

The future of water resource planning is one where **innovation, adaptability,** and **inclusivity** guide decision-making, and where traditional knowledge is celebrated as an integral part of the solution. Together, by building bridges between the **past and present**, we can create a future where **clean, accessible water** is available to all, and where water resource management becomes a collaborative, sustainable, and globally integrated effort.

Together, we can **redefine the future of water resource planning**—one that is **inclusive, innovative,** and **sustainable**. Let us create a **world** where **science** and **tradition** are not opposing forces, but **complementary allies,** working hand in hand to build a future of **water security, community resilience,** and **environmental sustainability.** The time to act is now—let us embrace this opportunity and build a water-secure world for generations to come.

◆ ◆ ◆

Appendices

MIND MAP-I

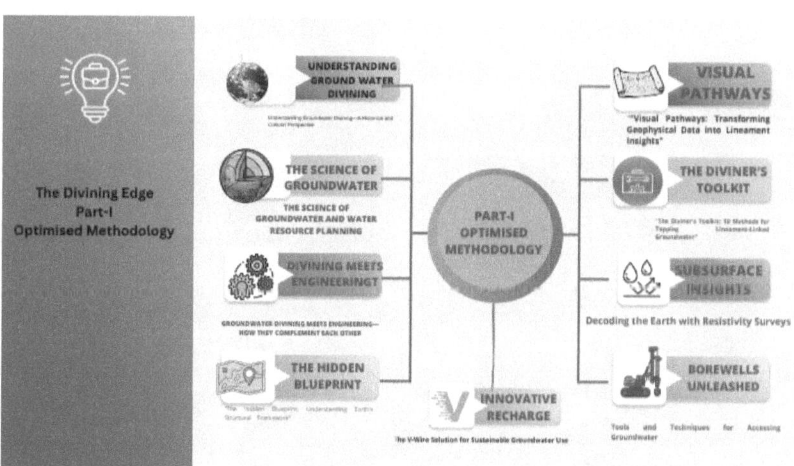

MIND MAP-II

PART-II, LEGIT GOALS
The Diviners Edge framework for Future implementation of Divining Technology
Traditional Divining Knowledge into modern water solutions
The Ethical and Practical considerations
Tools and techniques -A Practical Guide
Collaborating Engineers with Diviners and Case studies.
DIVINERS absolute Techniques
Part-I Optimised Methodology
PART-II LEGIT GOALS
Evaluating the Effectiveness
Diviners Metrics for Success, case studies and outcomes.
Innovative Model for Solutions
Overcoming Challenges
The Water Resource planning for sustainable Future
The Roadblocks to wider Adoption,capacity building and Regulatory

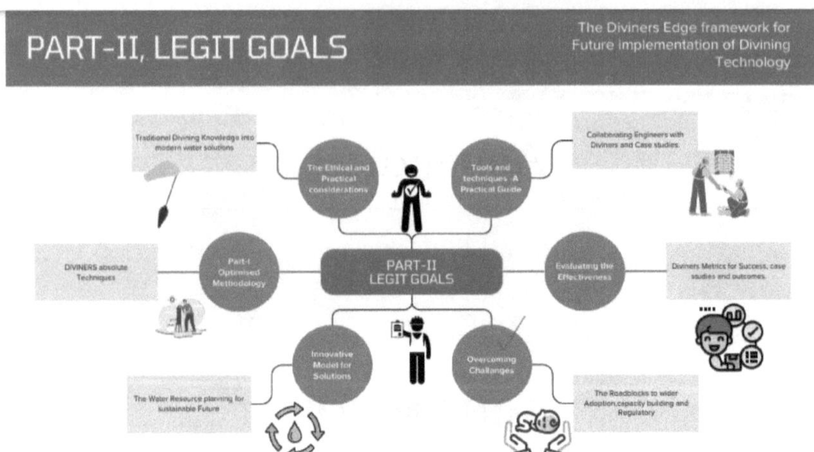

Resources and References

Books and Articles

1. Applied Remote Sensing for Urban Planning, Governance, and Sustainability

- Editors: MaikNetzband, William L. Stefanov, Charles Redman

- Focus: Remote sensing applications in urban and environmental studies, including geology.

2. Bhuvan Portal &Bhuvan User Handbook.

- Website: Indian platform for geospatial data.

3. Central Ground Water Board (CGWB) Reports

- Various publications on groundwater availability, management, and drilling guidelines in India.

- Website: www.cgwb.gov.in

4. "Dowsing and Water Divining: A Rational Analysis" by Reginald C. Knight

- Analyses the historical and cultural context of water divining and its modern applications.

5. **"Fundamentals of Geological Maps" by Richard J. Lisle**

   - A practical guide for interpreting geological maps.

6. **Geographic Information Systems and Science**

   - Authors: Paul A. Longley, Michael F. Goodchild, David J. Maguire, David W. Rhind

   - Description: A foundational book covering GIS principles and applications.

7. **"Geophysical Methods for Groundwater Exploration" by M. E. Schindel and K. F. Heinrich**

   - An in-depth guide to geophysical techniques used in groundwater exploration and borehole site selection

8. **Global Fault Database**

   - Website: Specialised in tectonic and fault data.

9. **Google Earth Engine**

   - Website: Ideal for processing and analysing large geospatial datasets.

10. **"Groundwater Hydrology" by David Keith Todd & Larry W. Mays**

    - Comprehensive coverage of groundwater science, including exploration, flow systems, and management practices.

11. **"Handbook of Applied Hydrology" by Vijay P. Singh**

    - A detailed guide to hydrological applications, including groundwater exploration and sustainable practices.

12. **"Handbook of Water Economics: Principles and Practice" by Colin Green**

13. **"Hydro-Dowsing: Ancient Practices and Modern Implications" by J. S. Michaelson (Journal of Cultural Geography, 2019)**

- A scholarly discussion on the intersection of cultural beliefs and practical applications of divining

14. **Hydrogeological Mapping in India: Concepts and Case Studies by R. Chatterjee**

- Focuses on hydrogeomorphological maps and groundwater prospecting specific to the Indian context.

15. **International Water Resources Association (IWRA)**

- Articles and research on sustainable groundwater management.

- Website: www.iwra.org

16. **Integration of Traditional and Scientific Methods in Water Resource Management" by J. Sharma & R. Singh (Journal of Hydrology, 2020)**

- Discusses how traditional practices like divining can complement modern hydrological techniques.

17. **LiDAR and Geophysical Data**

- High-resolution LiDAR data is valuable for detecting subtle topographic features that traditional remote sensing methods might miss.

18. **"Lineament Analysis in Geology" by A.K. Gupta**

- A specialised resource focused on lineament identification and interpretation using various datasets.

19. **Manual of Artificial Recharge of Groundwater by the Central Ground Water Board (CGWB), Government of India**

20. **Metrics and Models in Software Quality Engineering by Stephen H. Kan**

21. **My research paper on The Cultural Practice and Ethnoscience of Coconut Divining: 2025 (IWWA Raipur convocation)**

22. **Manual of RGNDW User Manual (India)**

23. **National Water Mission (NWM)**

- A portal for strategies and policies to ensure sustainable water use in India.

- Website: www.nwm.gov.in

24. **OpenTopography**

- Website: High-resolution topographic data.

25. **Rajasthan: Ecology and Culture by G.S. Bhalla**

26. **Remote Sensing and GIS**

- Lineament analysis using satellite imagery and Digital Elevation Models (DEMs).

27. **Respect for Cultural Knowledge in Sustainable Development edited by Gerald W. Creed**

28. **Rural Water Supply and Sanitation Guidelines by the Ministry of Jal Shakti, Government of India**

- Guidelines on rural water supply schemes and best practices for infrastructure development.

29. **Sentinel Hub EO Browser**

- Website: Access and visualise Sentinel satellite data.

30. **"The Cultural Politics of Water: A Critical Analysis of Contemporary Water Management Practices" by James Linton and Jessica Budds**

31. **"The Divining Edge: The Role and Importance of Diviners in Groundwater Search" by My research paper IWWA 2025 Raipur'**

32. **"The Water Diviner's Handbook: Harnessing the Wisdom of Water Dowsing" by Joe R. Chace**

33. UNESCO - Water Management Practices

- Reports on integrating traditional and scientific methods in water resource projects.
- Website: www.unesco.org

34. United Nations Water (UN-Water)

- Resources on traditional water knowledge and sustainable practices.
- Website: www.unwater.org

35. USGS Earth Explorer

- Website: Free access to Landsat and other datasets.

36. "Water Divining and Dowsing: An Exploration of Traditional Practices" by T. McLean

- A focused study on the historical and practical aspects of divining techniques.

37. WaterAid

- Resources on practical water solutions in rural communities, including traditional approaches.
- Website: www.wateraid.org

38. WaterAid India

- Practical guides on water resource management in rural and resource-poor areas.
- Website: www.wateraidindia.in

39. WaterAid India Resources

- Practical guides and research on water resource management in rural areas.
- Website: www.wateraidindia.in

40. "Water Well and Borehole Design" by F. Misstear, D. Banks, and A. Clark

- technical manual on modern borehole drilling practices and sustainable groundwater management.

◆ ◆ ◆

GLOSSARY OF TERMS

1. **2D/3D Resistivity Imaging:** Advanced resistivity techniques that create two- or three-dimensional models of subsurface resistivity, allowing for better resolution and interpretation in complex geological conditions.

2. **Aquifer**

   An underground layer of water-bearing rock or sediment where groundwater is stored and can be extracted through wells or boreholes.

3. **Artificial Intelligence (AI):** A field of computer science focused on creating systems that can perform tasks typically requiring human intelligence, such as recognising patterns, making decisions, or understanding natural language.

4. **Aspect Analysis**: Analysis of the direction that a terrain slope faces, helping in understanding geological orientation.

5. **Bhuvan Portal**: A geospatial platform developed by ISRO for accessing high-resolution satellite and terrain data, especially for India.

6. **Blasting for Tube Well Productivity:**The use of controlled explosive charges to create fractures or loosen dense geological formations to improve groundwater flow to a tube well.

7. **Borewell**

   A deep, narrow well drilled into the ground to access underground water resources, typically for domestic, agricultural, or industrial use.

8. **Cartosat**: A series of Indian satellites providing high-resolution imagery for mapping and terrain studies.

9. **Check Dams:**

   1. **Definition**: Small barriers built across watercourses to slow down water flow, enhance infiltration, and reduce erosion.

   2. **Function**: Promotes groundwater recharge and conserves surface water for dry seasons.

10. **Climate Change**

    Long-term changes in temperature, precipitation, and weather patterns caused by natural factors and human activities often exacerbate water scarcity issues.

11. **Collaboration**: Working jointly with others, especially in a multidisciplinary context, to achieve a common goal. In this context, it refers to engineers and diviners working together.

12. **Community-Driven Solutions**: Approaches to problem-solving that are initiated, led, and executed by the people directly affected by an issue, emphasising local input and ownership.

13. **Community Health**: The overall physical, mental, and social well-being of a population, often influenced by acccss to clean water.

14. **Community Resilience**

    The ability of a community to adapt and thrive in the face of challenges, such as water scarcity or environmental changes, is often supported by traditional practices.

15. **Community Satisfaction**: A measure of how well the needs, expectations, and preferences of local populations are met in the context of a project.

16. **Contrast Enhancement**: Increases visibility of subtle features in satellite imagery.

17. **Cost Savings:**

    The reduction in expenses is achieved by integrating dowsing methods with conventional engineering practices in water projects.

18. **Cross-Disciplinary Training**: Educational programmes that combine knowledge and skills from multiple fields or disciplines to address complex challenges.

19. **Cultural Alignment**

    The compatibility of project methods with local traditions, values and beliefs.

20. **Cultural Shift**: A significant change in the collective attitudes, practices, and norms of a society or group.

21. **Cultural Survival**

    The preservation of cultural practices, traditions, and knowledge systems that are essential to the identity and autonomy of communities.

22. **DEM (Digital Elevation Model)**: A digital representation of the Earth's surface topography used in terrain analysis.

23. **Dendritic Patterns**: Natural branching stream patterns, typically in homogenous bedrock.

24. **Divining (Water Divination):**

    A traditional method used to locate underground water is by interpreting signals through tools like L-Rods or Y-shaped branches believed to detect changes in electromagnetic fields.

25. **Drainage Pattern**: The arrangement of rivers and streams, which can be influenced by lineaments.

26. **Edge Detection**: Image processing techniques (e.g., Sobel, Canny filters) are used to highlight abrupt changes in image intensity, often corresponding to linear features.

27. **Ecological Knowledge:**

    Understanding of ecosystems and natural processes, often derived from traditional or scientific observations, is used to manage resources sustainably.

28. **Electromagnetic Field**

    A physical field produced by electrically charged objects, which diviners believe changes near water sources and are detectable using tools like L-Rods.

29. **Empirical Evidence:**

    Information and data obtained through observation, experimentation, and systematic research, forming the basis for scientific validation.

30. **Environmental Responsibility**

    Practices that minimise environmental damage, such as ensuring groundwater extraction does not lead to depletion or ecological harm.

31. **Exploratory Drilling**

    The initial phase of drilling small test wells to investigate subsurface conditions, such as groundwater availability, before full-scale extraction.

32. **Fault**: A fracture in the Earth's crust where rocks on either side have moved relative to each other.

33. **Fault Scarp**: A steep slope or cliff formed by movement along a fault.

34. **Flushing of Borewells; Definition**: A maintenance procedure that involves the removal of accumulated sediments, debris, or clogging materials from a borewell to restore water flow and quality.

35. **Fracture**: A break in rock formations that does not involve significant movement.

36. **Fractal Dimension**: A measure used to quantify the complexity of lineament patterns.

37. **Geographic Information Systems (GIS):** A framework for gathering, managing, and analysing spatial and geographic data. It allows users to visualise and interpret data to understand relationships, patterns, and trends in areas like water management.

38. **Geological Data**: Information about the physical composition and structure of the Earth, often used to identify potential water-bearing formations.

39. **Geological Map**: Represents rock types, fault lines, and structural features of an area, aiding in lineament identification.

40. **Geophysical Maps**: Represent subsurface data obtained from gravity and magnetic surveys.

41. **Geophysical Survey:**

    A method that uses electrical, magnetic or seismic technologies to identify underground formations and locate water sources.

42. **Geophysical Techniques:**

    Scientific methods, such as electrical resistivity and seismic reflection, are used to study underground formations and locate water sources accurately.

43. **GIS (Geographic Information System):** Software used to integrate and analyse spatial data layers (e.g., slope, geological maps) for mapping lineaments.

44. **Global Case Studies:** Real-world examples or research conducted in various regions to illustrate practical applications of specific methods or techniques.

45. **Gravity Anomaly:** Variation in gravitational field strength caused by subsurface density changes.

46. **Groundwater:**

    Water that collects beneath the Earth's surface in porous rocks and soil, forming an essential resource for drinking, agriculture, and industry.

47. **Groundwater Depletion:**

    1. **Definition:** The excessive extraction of groundwater resources beyond natural replenishment levels, leading to a decline in water tables.

    2. **Impacts:** Reduced agricultural productivity, water scarcity, and ecological imbalances.

48. **Groundwater Flow:**

    The movement of water through the subsurface, influenced by the permeability of rocks and the gradient of the land.

49. **Groundwater Mapping:** The process of identifying and delineating underground water sources using various methods, including geophysical surveys, satellite data, and traditional techniques such as dowsing.

50. **Groundwater Recharge:**

    The process by which water from rain or other sources seeps into the ground to replenish underground aquifers.

51. **Hillshade**: A shaded relief map that simulates light and shadows on a terrain surface to highlight elevation changes.

52. **Holistic Approach:**

    A strategy that considers multiple perspectives and integrates traditional knowledge with modern techniques to achieve comprehensive and effective outcomes.

53. **Hough Transform**: A feature extraction technique used to detect lines, curves, or other shapes in images.

54. **Hydrofracturing of Tube Wells Definition**: A technique that involves injecting water at high pressure into subsurface rock formations to create or expand fractures, thereby enhancing water flow to the well.

55. **Hydrogeological Mapping:**

    The creation of detailed maps to show groundwater distribution, aquifer boundaries, and water quality aids in resource planning and management.

56. **Hydrogeological Surveys**: Scientific investigations that study the distribution and movement of groundwater within the Earth's crust, often used in water resource projects.

57. **Hydrogeology:**

    A branch of geology that focuses on the distribution and movement of groundwater in rocks and soils.

58. **Hydrogeomorphological (HGM) Maps**: Combine topography, hydrology, and geomorphology to analyse groundwater flow and surface features influenced by lineaments.

59. **Hydrologist:**

A scientist or expert specialising in the study and management of water resources, including groundwater, surface water, and hydrological processes.

60. **Hydrology:**

The scientific study of water, including its distribution, movement, and interaction with the environment, both above and below the ground.

61. **Inclusivity:**

The practice of involving and valuing all individuals and perspectives, ensuring equitable participation in decision-making and resource distribution.

62. **Indigenous Practices:**

Traditional methods and techniques used by Indigenous communities, often rooted in cultural and ecological knowledge passed down through generations.

63. **InSAR (Interferometric Synthetic Aperture Radar):** A remote sensing technique using radar data to measure ground deformation.

64. **Institutional Resistance**: Hesitance or refusal by organisations or authorities to adopt new practices or ideas, often due to cultural, structural, or political barriers.

65. **Integrated Approach**: A strategy that combines multiple methods or disciplines to address a problem more effectively.

66. **Integrated Water Resources Management (IWRM):** A process that promotes the coordinated development and management of water, land, and related resources to maximise the economic and social welfare of a society while ensuring the sustainability of vital ecosystems.

67. 67. **Integration**: The process of combining different methods or systems to work together effectively, in this case, divining and engineering in borehole projects.

68. **Injection Wells:**

    1. **Definition**: Wells designed to inject water directly into aquifers to enhance groundwater storage.

    2. **Applications**: Used in managed aquifer recharge (MAR) projects and sustainable water management.

69. **Lessons Learned**: Insights or knowledge gained from the successes and failures of a project, which can inform future efforts.

70. **LiDAR (Light Detection and Ranging)**: A remote sensing method producing high-resolution 3D models of terrain, essential for identifying subtle lineaments.

    1. **DEM (Digital Elevation Model)**: Elevation data used for slope and hillshade analyses.

71. **Lineament**: A significant linear feature on the Earth's surface, representing geological structures such as faults, fractures, or folds.

72. **Local Knowledge**: Indigenous or community-based understanding of an area, developed through experience and tradition.

73. **L-Rods:**

    L-shaped metal rods used in dowsing to sense the presence of underground water or other minerals. The rods are held loosely in each hand and are thought to move or cross when water is detected.

74. **Magnetic Anomaly**: Variations in Earth's magnetic field related to subsurface structures.

75. **Machine Learning (ML):** A subset of AI that involves algorithms and statistical models enabling computers to learn from and make predictions based on data.

76. **Manual Drilling:**

    A traditional method of borehole construction involving hand-operated tools, often used in resource-poor areas.

77. **Mechanical Drilling:**

    The use of machinery to drill boreholes offers greater precision and the ability to access deeper water reserves.

78. **Metrics for Success**: Quantitative and qualitative measures used to evaluate the effectiveness and outcomes of a project or intervention.

79. **Modern Engineering Techniques:**

    Scientific methods and tools used for groundwater exploration, such as geophysical surveys, hydrogeological mapping, and mechanical drilling.

80. **Multispectral Imagery**: Satellite imagery capturing data across multiple wavelengths, useful for analysing different surface materials.

81. **NDVI (Normalised Difference Vegetation Index):** A spectral index used to measure vegetation health, sometimes useful for detecting subtle terrain features.

82. **NGOs (Non-Governmental Organisations):**

Independent organisations that operate without government control, often focused on humanitarian, environmental, or developmental goals.

83. **Overhead Tank:**

A raised water storage tank is used to distribute water via gravity to surrounding areas. This is commonly part of water supply systems in rural and urban infrastructure.

84. **Participatory Approach:**

   1. **Definition**: A method that involves local communities, stakeholders, and agencies in planning, implementing, and monitoring water resource management projects.

   2. **Significance:** Ensures ownership, sustainability, and effectiveness of interventions.

85. **Percolation Tanks:**

   1. **Definition**: Large, shallow tanks constructed to store rainwater and promote infiltration into the ground for groundwater recharge.

   2. **Benefits:** Improves water availability in surrounding areas and supports agriculture during dry periods.

86. **Policy Frameworks**: Structured approaches developed by governments or organisations to address specific issues, guiding decision-making and implementation.

87. **Proximity:**

The physical closeness or distance of one object or location to another, often considered in infrastructure planning for efficiency and cost-effectiveness.

88. **Public-Private Partnerships (PPP):** Collaborative agreements between government entities and private companies to finance, build, and operate services and infrastructure, such as water systems.

89. **Radiometric Correction:** Adjusting image data to remove sensor or environmental distortions for accurate analysis.

90. **Rectilinear Patterns:** Streams following straight paths, often fault-controlled.

91. **Regulatory Standards:** Official rules or guidelines established by authorities to ensure safety, efficiency, and fairness in various practices.

92. **Remote Sensing Technologies:** Methods of gathering data about the Earth's surface without direct contact, often through satellites or aerial imagery, to assess potential water sources. The use of satellite or airborne sensor technologies to gather information about the Earth's surface. In water management, it can be used to monitor water bodies, soil moisture, and drought conditions.

93. **Resistivity survey:** A geophysical method used to measure the subsurface electrical resistivity distribution by injecting electrical current into the ground and measuring the resulting potential difference.

94. **Rural Water Infrastructure:**

    Systems and facilities in rural areas are designed to provide access to water resources, including boreholes, overhead tanks, pipelines, and pumping mechanisms.

95. **Satellite Hydrology:** The use of satellite-based technology to study and manage water resources, especially in regions that are difficult to access with traditional methods.

96. **SAR/InSAR (Synthetic Aperture Radar/Interferometric SAR)**: Radar-based remote sensing techniques to detect surface displacement and identify active faults.

97. **Schlumberger Array:** A resistivity survey method where the current electrodes are spread further apart than the potential electrodes, enabling deeper subsurface investigations.

98. **Scientific Methods:**

Systematic processes used for inquiry and experimentation to test hypotheses and derive conclusions based on evidence.

99. **Scientific Scepticism**: A critical approach to evaluating claims, requiring empirical evidence and logical reasoning.

100. **Self-Sufficiency:**

The ability of a community or individual to meet their basic needs, such as water, independently and sustainably, often through localised practices like divining.

101. **Site Selection:**

The process of identifying and evaluating locations for specific purposes, such as drilling boreholes, based on technical, environmental, and logistical criteria.

102. **Slickensides**: Polished rock surfaces indicating fault movement.

103. **Slope Analysis**: A technique to measure terrain steepness, often used to identify structural features.

104. **Slope Map:** Highlights changes in elevation and steepness, useful for detecting faults or scarps.

105. **SRTM (Shuttle Radar Topography Mission)**: A global elevation dataset derived from radar measurements during NASA's 2000 mission.

106. **Statistical Comparisons**: The use of quantitative data to analyse and contrast outcomes between different approaches or scenarios.

107. **Sub-Saharan Africa**: A geographical region of Africa located south of the Sahara Desert, often used in discussions of development and resource management challenges.

108. **Success Rate:**

    The proportion of borehole projects that achieve their intended outcome, such as finding a reliable water source.

109. **Sustainability**: The capacity to use resources in a way that ensures their availability for future generations while minimising environmental impact.

110. **Sustainable Development Goals (SDGs):** Global goals set by the United Nations to address the world's most pressing issues, including clean water and sanitation (SDG 6).

111. **Sustainable Water Source:**

    A water source that can meet current needs without depleting or degrading the resource, ensuring availability for future generations.

112. **Synergy:**

    The combined effect of different methods or approaches that is greater than the sum of their individual contributions, as seen in integrating divining with modern engineering.

113. **Time and Cost Savings**: Reductions in the resources required to complete a project, achieved through efficient methods or practices.

114. **Time Reductions:**

    The decrease in the time required to locate and access water sources due to the efficiency of dowsing techniques.

115. **Topographic Map**: A detailed map showing the terrain's contours and elevation, often used to identify features like ridges, valleys, and slopes.

    1. **Slope Map**: Highlights changes in elevation and steepness, useful for detecting faults or scarps.

    2. **Aspect Map**: Indicates the direction of slopes, aiding in identifying linear geological patterns.

116. **Tourism Motel:**

    A lodging facility designed to cater to tourists, typically offering basic amenities and located in or near areas of interest.

117. **Traditional Knowledge:**

    Knowledge, practices and techniques developed over generations, often based on cultural beliefs and empirical observations, such as the use of divining for water location.

118. **V-Wire Technology:**

    1. **Definition**: A type of well screen made from wedge-shaped wires, offering efficient water filtration and durability. Used in groundwater recharge systems to filter and manage wastewater before infiltration into aquifers.

    2. **Applications**: Effective in rural water management projects for filtering and recharging groundwater using wastewater.

    3. **Key Components:**

        1. **Filtration**: Removes sediments and contaminants.

        2. **Recharge**: Promotes aquifer replenishment.

        3. **Sustainability**: Reduces groundwater depletion and supports water security.

119. **Vertical Electrical Sounding:** A geophysical method involving resistivity measurements at various depths by increasing the distance between electrodes. It is widely used for determining the depth and properties of aquifers.

120. **Water Demand:**

The quantity of water required by a population or industry for drinking, agriculture, or industrial processes is often influenced by population growth and climate conditions.

121. **Water Management:**

The process of planning, developing, distributing, and optimising water resources to meet societal and environmental needs.

122. **Water Resource Management:**

The process of planning, developing and managing water resources to ensure sustainable and efficient use for domestic, industrial, agricultural and environmental purposes.

123. **Water Scarcity:**

A condition where water demand exceeds the available supply in a specific region, often caused by natural or human factors such as climate change, overuse, or mismanagement.

124. **Water Security:** The availability of sufficient and clean water to meet the needs of communities, agriculture, industry, and ecosystems without jeopardising future generations' access to water.

125. **Water Yield:**

The amount of water that can be extracted from a borehole or other water source over a specific period, often measured in litres per minute or cubic metres per hour.

126. **Watershed Management:**

  1. **Definition:** The process of conserving and managing water resources in a specific drainage area to support sustainable livelihoods and ecosystems.

  2. **Components:** Includes measures such as soil conservation, afforestation, and water harvesting structures.

127. **Wenner Array:** A configuration of electrodes used in resistivity surveys where four electrodes are spaced equally in a straight line to measure subsurface resistivity. It is ideal for shallow investigations due to its simplicity and ease of use.

◆ ◆ ◆

Index

Thank You